María E. Otegui
Gustavo A. Slafer
Editors

Physiological Bases for Maize Improvement

Pre-publication
REVIEWS,
COMMENTARIES,
EVALUATIONS . . .

"**P**hysiological Bases for Maize Improvement* edited by Otegui and Slafer is a needed addition to literature about the important aspects of maize yield improvement worldwide and most factors affecting maize performance under realistic field conditions. A group of leading authorities contributed state-of-the-art scientific information regarding several aspects of maize improvement in temperate and tropical climates as well as under water deficit conditions."

Joe T. Ritchie
Professor of Crop and Soil Sciences,
Homer Nowlin Chair,
Michigan State University,
East Lansing

Food Products Press®
An Imprint of The Haworth Press, Inc.

Physiological Bases for Maize Improvement

FOOD PRODUCTS PRESS
Crop Science
Amarjit S. Basra, PhD
Senior Editor

New, Recent, and Forthcoming Titles of Related Interest:

Dictionary of Plant Genetics and Molecular Biology by Gurbachan S. Miglani

Advances in Hemp Research by Paolo Ranalli

Wheat: Ecology and Physiology of Yield Determination by Emilio H. Satorre and Gustavo A. Slafer

Mineral Nutrition of Crops: Fundamental Mechanisms and Implications by Zdenko Rengel

Conservation Tillage in U.S. Agriculture: Environmental, Economic, and Policy Issues by Noel D. Uri

Cotton Fibers: Developmental Biology, Quality Improvement, and Textile Processing edited by Amarjit S. Basra

Heterosis and Hybrid Seed Production in Agronomic Crops edited by Amarjit S. Basra

Intensive Cropping: Efficient Use of Water, Nutrients, and Tillage by S. S. Prihar, P. R. Gajri, D. K. Benbi, and V. K. Arora

Physiological Bases for Maize Improvement edited by María E. Otegui and Gustavo A. Slafer

Plant Growth Regulators in Agriculture and Horticulture: Their Role and Commercial Uses edited by Amarjit S. Basra

Crop Responses and Adaptations to Temperature Stress: New Insights and Approaches edited by Amarjit S. Basra

Physiological Bases for Maize Improvement

María E. Otegui
Gustavo A. Slafer
Editors

Food Products Press®
An Imprint of The Haworth Press, Inc.
New York • London • Oxford

Published by

Food Products Press®, an imprint of The Haworth Press, Inc., 10 Alice Street, Binghamton, NY 13904-1580

Cover design by Jennifer Gaska.

Workshop logo on front cover designed by Paula Otegui.

Library of Congress Cataloging-in-Publication Data

Physiological bases for maize improvement / Maria Elena Otegui, Gustavo A. Slafer, editors.
    p. cm.
  Includes bibliographical references and index.
  ISBN 1-56022-889-X (alk. paper)
    1. Corn—Yields. 2. Corn—Physiology. 3. Crop improvement. I. Otegui, Maria Elena. II. Slafer, Gustavo A., 1960-

SB191.M2 P49 2000
633.1′558—dc21
                                      00-022372

CONTENTS

ABOUT THE EDITORS

María E. Otegui, DrSci, is Adjunct Professor of Grain Crops Production and a member of the staff of the School for Graduate Studies at the University of Buenos Aires in Argentina. Dr. Otegui is a member of the National Council for Scientific and Technical Research (CONICET) in Argentina, the Crop Science Society of America, the American Society of Agronomy, the Argentine Society of Plant Physiology, and the Argentine Society of Genetics.

Gustavo A. Slafer, PhD, is Associate Professor of Grain Crops Production and Chair of the Plant Production Program of the School for Graduate Studies at the University of Buenos Aires in Argentina. Dr. Slafer is editor of the book *Genetic Improvement of Field Crops* and co-editor of the book *Wheat: Ecology and Physiology of Yield Determination* (Haworth). He is an Editorial Board member of the *Journal of Crop Production* (Haworth) and speaks internationally on crop physiology, particularly in relation to yield generation and breeding. He is a member of the National Council for Scientific and Technical Research (CONICET) in Argentina, the Institute of Physiology and Ecology Related to Agriculture, the Crop Science Society of America, the American Society of Agronomy, the Argentine Society of Plant Physiology, and the Argentine Society of Genetics, among others.

Contributors

Fernando H. Andrade, PhD, is Professor, Unidad Integrada Instituto Nacional de Tecnología Agropecuaria (INTA) y Facultad de Ciencias Agrarias de la Universidad Nacional de Mar del Plata, Balcarce, Argentina.

Marianne Bänziger, PhD, is Maize Physiologist, Centro Internacional de Mejoramiento de Maíz y Trigo (CIMMYT), Harare, Zimbabwe.

Jorgelina Cárcova, MSc, is Research and Teaching Agronomist, Departamento de Producción Vegetal, Facultad de Agronomía, Universidad de Buenos Aires, Buenos Aires, Argentina.

Alfredo G. Cirilo, Dr, is Researcher, Instituto Nacional de Tecnología Agropecuaria (INTA), Pergamino, Argentina.

Laura Echarte, Agr. Eng., is a doctorate student, Unidad Integrada Instituto Nacional de Tecnología Agropecuaria (INTA) y Facultad de Ciencias Agrarias de la Universidad Nacional de Mar del Plata, Balcarce, Argentina.

Gregory O. Edmeades, PhD, is Researcher, Pioneer Hi-Bred International, Incorporated, Kekaha, Hawaii.

Claudio M. Ghersa, Agr. Eng., is Associate Professor, Departamento de Fisiología y Ecología Vegetal, Facultad de Agronomía, Universidad de Buenos Aires, Buenos Aires, Argentina.

Anthony Hunt, PhD, is Professor, Department of Plant Agriculture, Crop Science Division, University of Guelph, Guelph, Ontario, Canada.

James R. Kiniry, PhD, is Research Agronomist, U.S. Department of Agriculture (USDA), Temple, Texas.

Gustavo A. Maddonni, MSc, is Research and Teaching Agronomist, Departamento de Producción Vegetal, Facultad de Agronomía, Universidad de Buenos Aires, Buenos Aires, Argentina.

Russell C. Muchow, PhD, is Research Leader, Commonwealth Scientific and Industrial Research Organization (CSIRO), Division of Tropical Agriculture, and CRC for Sustainable Sugar Production, Cunningham Laboratory, St. Lucia, Brisbane, Australia.

Jean-Marcel Ribaut, PhD, is Molecular Geneticist, Centro Internacional de Mejoramiento de Maíz y Trigo, Mexico DF, Mexico.

Víctor O. Sadras, PhD, is Professor, Unidad Integrada Instituto Nacional de Tecnología Agropecuaria (INTA) y Facultad de Ciencias Agrarias de la Universidad Nacional de Mar del Plata, Balcarce, Argentina.

Matthijs Tollenaar, PhD, is Professor and Department Head, Department of Plant Agriculture, Crop Science Division, University of Guelph, Guelph, Ontario, Canada.

Mark E. Westgate, PhD, is Associate Professor, Department of Agronomy, Iowa State University, Ames, Iowa.

Jiangang Wu, MSc, is Graduate Research Assistant, Department of Plant Agriculture, Crop Science Division, University of Guelph, Guelph, Ontario, Canada.

Preface

Although isolated examples of the physiological bases of genetic improvement of maize yield may be found in several papers (many included as chapters in this book), a single volume does not appear to condense the available information in this field. Most studies quantifying increases in maize yields associated with its genetic improvement reported breeders' remarkable success. However, as the population keeps burgeoning and the increases in agricultural inputs appear to slow down, the satisfaction of the expected demand in the new century would strongly (even more so than during the past century) depend upon genetic gains. In other words, though empirical selection for maize yield may continue to be effective, genetic gains, even higher than the relatively high gains obtained in the past few decades, together with the likelihood of using new technologies (such as molecular biology), demand a better understanding of crop physiology to be effectively and efficiently used in crop breeding. With this in mind, we organized an international workshop, which took place in 1998 (October 8-9) in Buenos Aires, whose principal aim was to join the views and ideas of some of the most recognized scientists in this area to define prospective physiological characteristics that may be useful in realistic breeding programs. In this context, the primary aim of the workshop, and of this book, was to bring together the most recent research on physiological bases that may be useful in maize improvement. The original presentations made by the speakers at this workshop were the basis of this volume, but, different from other books reporting the results from workshops (or other scientific meetings), these presentations were rewritten after the workshop ended to include the gains that each of us made on that occasion.

This book attempts to provide a physiological approach to understanding the generation of maize yield that might be useful in future breeding. All authors attempted to bring together the relevant information dispersed among many different journals, providing an idea of the state of the art in each particular field, intending their chapters to be fair and critical updates of the knowledge developed, particularly in the last decade. The book has eleven chapters, with the majority focusing on the physiological bases of maize breeding for higher yield potential and for tolerance to biotic and physical stresses (Chapters 2 through 9). These chapters are preceded by an introductory chapter on the reasons why we believe there is a role for physiology in future maize breeding (Chapter 1). The book concludes with a chapter on the use of models to assist breeding (Chapter 10) and one chapter that synthesizes much of the workshop discussions, based on selected poster presentations on yield potential and stress tolerance (Chapter 11).

The book is mainly addressed to breeders and physiologists working in different fields related to maize yield, its determining processes, and how they may be used in breeding higher-yielding hybrids. The audience also includes breeders and physiologists interested in other crops with similar situations (most major crops) as well as other professionals related to this field (e.g., agronomists, crop ecologists, entomologists, weed scientists). Based on the prestige and well-gained worldwide recognition of most authors, this book will become, we trust, a useful reference for advanced undergraduate and postgraduate students in courses on plant/crop physiology, plant breeding, crop production, and other related subjects.

As frequently occurs in multiauthored books, some repetition among chapters, as well as occasional disagreement among contributors, emerges. We have decided to leave these qualities as a strategy to highlight similarities and differences in the perspectives of the authors.

All authors were quite happy to contribute their ideas from the workshop and, afterward, to make the extra effort to produce their chapters, and we would like to express our sincere gratitude to each of them. We also thank our colleagues in the Cereal Production Unit, within the Department of Plant Production, Faculty of Agronomy, University of Buenos Aires, in particular, the scientists working in our labs (R. Savin, D. J. Miralles, G. Maddonni, D. F. Calderini, J. Cárcova, M. F. Dreccer, L. G. Abeledo, E. Whitechurch, L. Borrás, and M. Uribelarrea), who understood the extra time demanded, first, by the organization of the workshop and, second, by the editing of this book.

We would like to thank, most of all, Fundación Antorchas from Argentina, which was the original funding body that supported this project, and without which nothing would have been initiated. Other Argentine or multinational organizations supported the workshop further and allowed us to invite more scientists to participate in it; they were Dekalb Argentina, Asociación de Semilleros Argentinos, Nidera, and Morgan-Mycogen, to which we are also thankful. We are also grateful to the U.S. Department of Agriculture, which supported the workshop as well by covering the airfare of one participant. We would also like to acknowledge the help received from the authorities of our faculty and the administration of the School for Graduate Studies in organizing the workshop, particularly Jorgelina Cárcova and Lucas Borrás for their assistance during the workshop.

Despite the time required, we have truly enjoyed both organizing/ hosting the workshop and editing this book. We hope readers will find it (as a whole or in part, depending on their interests) useful and stimulating.

María E. Otegui
Gustavo A. Slafer

Chapter 1

Is There a Niche for Physiology in Future Genetic Improvement of Maize Yields?

Gustavo A. Slafer
María E. Otegui

WORLDWIDE IMPORTANCE OF MAIZE

From the almost quarter of a million recognized plant species in the world, only very few (less than 100 per million) are used as crops to satisfy virtually all the human requirements for food and fiber (Wittwer, 1980). As our future welfare may strongly depend upon the productivity of these few plant species, we need always to improve our understanding about their potential productivity and adaptability to environmental constraints (Evans, 1975).

Among these few species, the cereals are, by severalfold, the most important, with a global production of over 2 billion tons during recent years (Food and Agriculture Organization [FAO], 1998). As a simply comparative measure of the relative importance of cereals, the global production of pulses has been around 55.5 million tons (more than thirty-five times less than the production level of cereals).

Regarding their annual production volumes, wheat, maize, and rice, in that order (but with very close figures, of around 570 million tons per year), are the main crops in the world, separated by a huge difference from the fourth- and fifth-ranked crops (barley and soybeans, with about 150 and 135 million tons per year, respectively). Despite maize being higher yielding than wheat and rice, considering its total area sown per year also provides an impressive measure of

1

maize world importance. Of worldwide arable land, 10 percent of just less than 1.4 billion hectares (ha) (20 percent of worldwide land sown to cereals) has been sown to maize during recent decades (see Figure 1.1). Despite its importance, there has been no evidence of clear changes in the area in which maize (and other cereals) is grown, and, thus, any changes in production were almost exclusively due to parallel changes in yield. On the other hand, energy devoted to agriculture, in general (e.g., see use of fertilizers and number of tractors in Figure 1.1), increased by severalfold during these decades, with a much reduced rate in the 1990s. Although these figures are for agriculture in general, maize has not been an exception. In the United States, for example, nitrogen (N) fertilizers were rarely used until

FIGURE 1.1. Worldwide Area Sown to Cereals (Open Squares) and to Maize (Open Circles) Since 1961

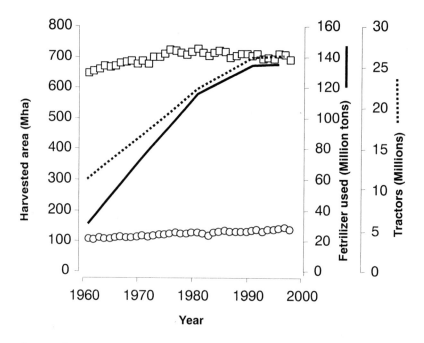

Source: Raw data taken from the FAO statistical databases.

hybrids became widespread, and from then (around the mid-1940s) to the early 1970s, the rate increased to about 130 kilograms (kg) of N per ha. A similar pattern can be seen in the use of herbicides (see Figure 22 in Evans, 1998).

TRENDS IN POPULATION GROWTH AND MAIZE YIELDS

A clear description of trends in population throughout the history of humankind on Earth, particularly in relation to agriculture evolution, has been recently offered by Evans (1998). In this chapter, we concentrate our attention only on the relatively short window of approximately a century, preceding a new millennium. World population has been increasing at the very high rate of almost 2 percent since the mid-1900s, determining a change from 2.5 billion people in 1950 to approximately 6 billion at the end of the century. Although this increase in population is striking (more than doubling its figure in less than fifty years, while it took more than 99 percent of humankind's time on Earth to reach a population of only 5 to 15 million, and almost 10,000 years to reach the first billion by the first quarter of the 1800s; see Rasmuson and Zetterström, 1992, and references in Evans, 1998), food production has also more than doubled during the last half of the twentieth century (Byrnes and Bumb, 1998). In most agricultural products, including maize, these changes were associated with increases in yields, with no clear trends in the area under production after the mid-1900s (for maize and other cereals, see Figure 1.1).

Maize yields, averaged across all maize growing areas in the world, have been steadily growing during the last decades, at the remarkable average rate of slightly more than 60 kg · ha^{-1} per year. This trend is particularly important if we consider that (1) it is much stronger than that corresponding to previous decades of the present century (for example, virtually no maize yield gains were seen in the United States for the initial four decades of the present century; see Figure 22 in Evans, 1998), and (2) it does not appear to exhibit any sign of having reached a ceiling. The second aspect shows a difference with wheat yield trends for recent years averaged across the world (Slafer, Calderini, and Miralles, 1996), as well as for many of the producing countries (Calderini and Slafer, 1998). In fact, during the past four decades (taken as a unique period), maize yields grew proportionally more than the world population (the population increased 100 percent; maize yields grew 220 percent; see Figure 1.2a, inset), despite the troubling increase in population experienced during the second half of this century (see previous discussion). However, although the linear regression of maize yields against years (1961-1997) was significant, two trends can be distinguished by splitting the whole period at the beginning of the 1980s, with much higher yield gains during the initial (1961-1980) than the latest (1981-1997) phase (Figure 1.2a). Although it will not be analyzed further here, it is also remarkable that yields, on a global basis, appear to have been losing stability as well, which is in agreement with a previous analysis made for maize yield trends in Australia (Slafer and Kernich, 1996).

As population growth has not slowed at all, observing the growth of maize yield and population after 1980, symptoms that we may be facing troubles in keeping yield gains at the pace of world population growth in the near future emerge at first sight (Figure 1.2b). In addition, yield gains shown in Figure 1.2 reflect the net result of those produced by both genetic and management improvements. Even accepting that yield increases matched the population trend in the past (which is not true for the past fifteen to twenty years), a change in the expected genetic gains for the future will be needed. This is because, due to economic and environmental reasons, further increases in yields through management gains will be harder to achieve than in the past (the clear decline in the use of some inputs during recent decades may be taken as a symptom of this; see Figure 1.1). Therefore, the contribution of maize breeding to yield increases in the future should not be maintained but increased, to replace as much as possible the management gains based on greater inputs that are likely to be dramatically reduced. In other words, even if maize yields were keeping pace with population growth, its genetic improvement should be more efficient than in the past, as yield gains would increasingly rely on breeding. However, since maize yields during the past fifteen to twenty years actually have been failing to keep pace with population increase (Figure 1.2a, inset), the effort must be even greater.

FIGURE 1.2. World Average Yields for Maize Crops Since 1961

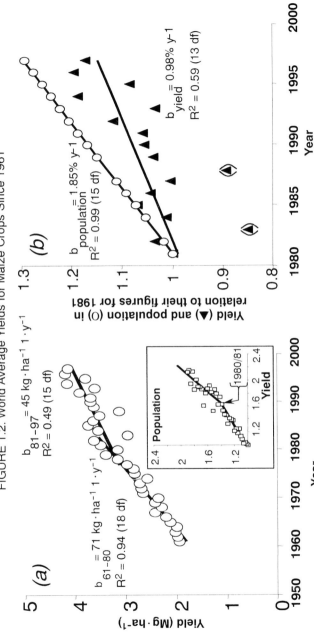

Source: Raw data taken from the FAO Web site (www.fao.org) (1999).

Note: (*a*) Line fitted by linear regression. The relationship between world population and yield for the 1961-1997 period, in relative values to those corresponding to 1961, is shown inset (lines fitted by eye); and, in more detail, that between their relative values to those corresponding to 1981, with lines fitted by regression, excluding data between parentheses (*b*). df = degrees of freedom.

Finally, the increase in maize yields observed in recent years on a worldwide basis has not been that clear for all maize regions (see Figure 1.3). While some countries (e.g., France and Argentina) exhibited a sustained increase in yields, others have shown a less evident trend to increase yields (e.g., the United States and Canada), and still others virtually did not increase their yields during the 1990s (e.g., China and Democratic Republic of the Congo). In fact, the coefficient of determination of the linear regressions shown in Figure 1.3 (ranging from 0.01 to 0.86) was significant ($P < 0.05$) for only eight of the twenty-five analyzed countries.

GENETIC CONTRIBUTIONS TO GAINS IN MAIZE YIELDS

Maize has been continuously under selection, at least by "natural" selection of landraces in ancient agriculture (by environmental factors and competition as well as likely conscious selection made by farmers, e.g., choosing the largest ears as a source of seed for the following crop). The greatest genetic gains in yield undoubtedly occurred, however, after the rediscovery of Mendel's laws. During this period of scientifically based breeding, the most dramatic changes in yield progress were associated with the introduction of hybrids, first, double cross and, more recently, single cross. In Iowa, for example, yield gains occurring during the periods of cropping varieties (prior to the 1930s), double-cross hybrids (1930s to 1960s), and single-cross hybrids (since 1960s) were equivalent to approximately 1, 55, and 95 kg $\cdot$ ha^{-1} per year (Troyer, 1990), respectively. Although many other technological improvements (than the simple introduction of "better-performing genetic material") have also been playing a role (Evans, 1998), it appears that replacement of open-pollinated varieties by double-cross and then by single-cross hybrids has been, undoubtedly, the driving force behind the impressive change in yield gains during this century. Tollenaar and colleagues (1994) succinctly described the "history" of maize breeding for hybrids (in the United States). They showed that it took virtually thirty years from the beginning of hybrid breeding to the wide adoption of the double-cross hybrids, and another three decades or so for the wide diffusion of the single-cross ones (see also Russell, 1991).

FIGURE 1.3. Average National Yields for Maize Crops from Many Different Countries, 1990 to 1997

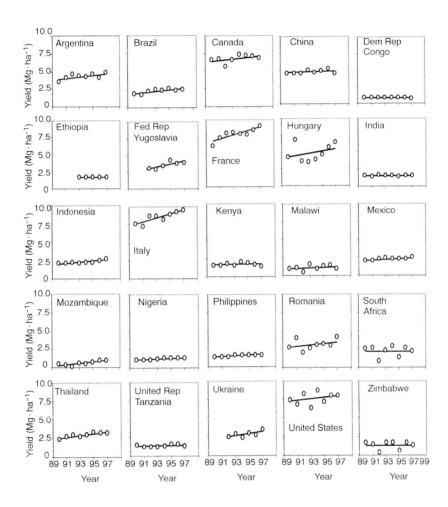

Source: Raw data taken from the FAO Web site (www.fao.org) (1999).

Note: Lines fitted by linear regression. Countries included in this figure are all those sowing maize on more than 1 million hectares per year (average of the last three years).

Although several reasons make it difficult to separate the genetic gains in yield from other contributions (Harper, 1983; Slafer and Andrade, 1991), when looking for future demands that maize production may have on breeding, it is relevant to estimate the relative contribution of genetics to the total gain in yield (increases due to both genetics and management). Russell (1991) compiled several sources in which the genetic contribution to yield gains in maize made by the release of new hybrids ranged from 29 to 89 percent of the total yield gains (see Table V in Russell, 1991). As far as we are aware, the latest calculation made (Duvick, 1992) reported an estimated genetic contribution to total gains in yield of about 55 percent for the period 1930 to 1990 in Iowa. However, the calculated contribution accepted a linear increase in maize yields throughout the analyzed period, from the release of the first hybrids to 1989. Although the linear component of the regression is undoubtedly large (as the R^2 was highly significant), it is clear that only a nonlinear sigmoid curve allows a random distribution of the residuals and, therefore, does better describe the trend in yield with the year of hybrid release (see Figure 1.4).

A simple and quick view of these results suggests that there was a stepwise advance in yielding capacity of maize by the introduction of the first hybrids (comparing the behavior of the oldest, lowest-yielding hybrids with the open-pollinated cultivar, Reid's Yellow Dent, which was very popular in the 1930s; see Figure 1.4). This is expected to occur in any crop in which varieties are replaced on a large scale by hybrids: the higher seed cost requires this replacement to take place fast; hence, a clear yield advantage can be expected. However, after the gain produced by breeding hybrids instead of open-pollinated cultivars, genetic gains were rather poor in the first decade or so of hybrid breeding, followed by an impressive efficiency in breeding hybrids of higher yielding ability during the period from midcentury to the end of the 1970s (see Figure 1.4). A clear linear increase in yield with the release of newer hybrids during these three decades was also shown by Russell (1991).

The efficiency of maize breeding since the end of the 1970s has apparently dropped quite substantially, however (see Figure 1.4), and this may be one of the reasons (together with the diminishing increases in energetic inputs discussed earlier) why actual yields in the

FIGURE 1.4. Grain Yields (0 Percent Moisture) of Maize Hybrids Released at Different Eras in the United States Plotted Against Their Year of Introduction

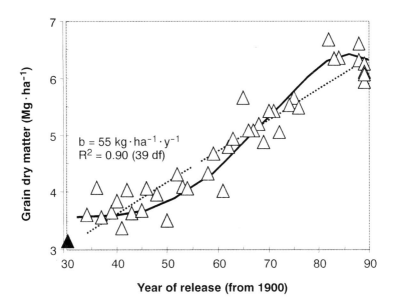

Source: Adapted from Table 3 in Duvick (1992, p. 73).

Note: Data include an open-pollinated cultivar released in 1930 (closed triangle). Dotted line was fitted by regression (parameters inset) and the sigmoid curve was fitted by eye. Each data point is the average of two years, three locations, and three planting densities.

United States since 1980 have not followed the upward momentum of the 1930 to 1980 period (see Figure 1 in Duvick, 1992; complemented by Figure 1.3, for the United States, in the present chapter, in which the lack of clear trends continues at least until 1997).

A NICHE FOR CROP PHYSIOLOGY IN MAIZE BREEDING

We are now more than sixty-five years apart from the major step in maize genetic improvement of breeding hybrids, and approximately three decades beyond the last large step in advancing yield gains by changing breeding hybrids from double- to single-cross. Genetic gains are apparently slowing down just when future demands of increased production would depend on them more than ever, and, as far as we are aware, there are no clear indications of what are the likely sources for the next step. Maize breeders face a crop with an already highly improved yield potential that needs to be further increased. Because the procedures that have been successfully used up to now will possibly not be as efficient in the future as they were in the past, genetic improvement in yield potential will be harder to obtain than it has been thus far.

Within the complex scenario of genetic gains being slightly reduced when breeding is expected to contribute more than ever to increased maize production, it may be expected that using physiological attributes to complement empirical selection could help in accelerating genetic improvement (Shorter, Lawn, and Hammer, 1991). In addition, as Slafer, Calderini, and Miralles (1996) suggested, the effectiveness of selection by physiological traits also could be recognized, in that it allows selection to be made during early generations, sometimes selecting for individual plant characteristics, and thus reducing the number of plots to be assessed and the size of the program, which otherwise would become progressively larger (Austin, 1993).

Molecular biology does certainly seem to be a promising tool when discussing breeding for qualitative traits, but things are less clear when targeting to increase productivity or tolerance to complex stresses. Although we recognize that some workers reported that they have identified quantitative trait loci (QTLs) for yield, water use efficiency, and other complex quantitative characteristics, the genotype $\times$ environment (G $\times$ E) interactions in the expression of these QTLs are frequently very large. Thus, it appears that we need to understand better the causes underlying those G $\times$ E interactions, through a collaboration between crop physiologists and molecular biologists, before these promising techniques can be widely used to increase yield potential and tolerance to complex and unpredictable stresses.

Different approaches identify which are the physiological attributes with potential to be used in breeding. Among other approaches, historical series of yield data have been largely used to determine which characteristics of the crop were modified when breeders successfully increased yield potential (see reviews by Russell, 1991; Tollenaar, McCullough, and Dwyer, 1994, and references in them). Alternatively, studies of the generation of yield components and their interrelationships under different environmental conditions may also help in the understanding of potentially useful traits to be targeted. Crop modeling will likely play a much more important role than it has played in the past in identifying characteristics to improve performance, particularly in variable environments in which the $G \times E$ interaction is large. We attempted to include in this book the authors' views on some of these alternatives.

Various yield-determining characteristics can be considered for future selection aimed at increasing yield potential. These include those related to the ability of the crop to efficiently intercept and use the available incident solar radiation, to allocate different proportions of its growth into competing organs, and to absorb and use water and nutrients. Most of the work done by the authors of the specific chapters addressed these topics, first, as tools to maximize yield potential under different environmental regimes, such as those characteristic of high latitude, temperate, and tropical regions, and, second, as physiological keys that modify the crop responses to environmental stresses, such as water, nutrients, pests, and weeds. We have also attempted to understand how these traits are identified and used in a breeding program, and how crop modeling can help in this process.

CONCLUDING REMARKS

During the present century (or at least for most of it), worldwide maize production has been dramatically increased, not only matching, but surpassing, population growth. This increase has been largely associated with changes in yield related to the change from using open-pollinated varieties to single-cross hybrids. Nevertheless, as current trends appear unable to keep pace with increases in a world population that keeps burgeoning, we must find alternatives to complement conventional breeding for yield per se. In this context, it may be helpful to increase our understanding of physiological-ecological processes at the crop level of organization that may help identify realistic opportunities for future breeding, either using these traits directly in breeding or improving the usefulness of molecular biology in contributing to breeding for quantitative complex traits.

In this book, we hope to have joined the views and ideas of recognized international scientists, who have faced different agroecological conditions for maize cropping, with the experience of breeders, physiologists, and agronomists. It was our aim to identify which prospect characteristics may be useful in realistic breeding programs and which traits deserve further in-depth study.

REFERENCES

Austin, R.B. (1993). Augmenting yield-based selection. In Hayward, M.D., Bose-mark, N.O., and Romagosa I. (Eds.), *Plant breeding: Principles and prospects* (pp. 391-405). London: Chapman and Hall.

Byrnes, B.H. and Bumb, B.L. (1998). Population growth, food production and nutrient requirements. In Rengel, Z. (Ed.), *Nutrient use in crop production* (pp. 1-27). Binghamton, NY: Food Products Press.

Calderini, D.F. and Slafer, G.A. (1998). Changes in yield and yield stability in wheat during the 20th century. *Field Crops Research* 57: 335-347.

Duvick, D.N. (1992). Genetic contributions to advances in yield of U.S. maize. *Maydica* 37: 69-79.

Evans, L.T. (1975). Crops and world food supply, crop evolution and the origins of crop physiology. In Evans, L.T. (Ed.), *Crop physiology* (pp. 1-22). Cambridge: Cambridge University Press.

Evans, L.T. (Ed.) (1998). *Feeding the ten billion: Plants and population growth.* Cambridge: Cambridge University Press.

Food and Agriculture Organization (FAO) (1998). *FAO Statistics.* Web site of the Food and Agriculture Organization of the United Nations, Rome, Italy. <www.fao.org>.

Harper, F. (Ed.) (1983). *Principles of arable crop production.* London: Granada Publishing Ltd.

Rasmuson, M. and Zetterström, R. (1992). World population, environment and energy demands. *Ambio* 21: 70-74.

Russell, W.A. (1991). Genetic improvement of maize yields. *Advances in Agronomy* 46: 245-298.

Shorter, R., Lawn, R.J., and Hammer, G.L. (1991). Improving genotypic adaptation in crops—A role for breeders, physiologists and modellers. *Experimental Agriculture* 27: 155-175.

Slafer, G.A. and Andrade, F.H. (1991). Changes in physiological attributes of the dry matter economy of bread wheat (*Triticum aestivum*) through genetic improvement of grain yield potential at different regions of the world: A review. *Euphytica* 58: 37-49.

Slafer, G.A., Calderini, D.F., and Miralles, D.J. (1996). Yield components and compensation in wheat: Opportunities for further increasing yield potential. In

Reynolds, M.P., Rajaram, S., and McNab, A. (Eds.), *Increasing yield potential in wheat: Breaking the barriers* (pp. 101-133). Mexico: CIMMYT.

Slafer, G.A. and Kernich, G.C. (1996). Have changes in yield (1900-1992) been accompanied by a decreased yield stability in Australia cereal production? *Australian Journal of Agricultural Research* 47: 323-334.

Tollenaar, M., McCullough, D.E., and Dwyer, L.M. (1994). Physiological basis of the genetic improvement of corn. In Slafer, G.A. (Ed.), *Genetic improvement of field crops* (pp. 183-236). New York: Marcel Dekker.

Troyer, A.F. (1990). A retrospective view of corn genetic resources. *Journal of Heredity* 81: 17-24.

Wittwer, S.H. (1980). The shape of the things to come. In Carlson, P.S. (Ed.), *The biology of crop productivity* (pp. 413-459). New York: Academic Press.

Chapter 2

Improving Maize Grain Yield Potential in a Cool Environment

Matthijs Tollenaar
Jiangang Wu

INTRODUCTION

A retrospective analysis of the physiological basis of yield improvement may provide an understanding of yield potential and indicate avenues for future yield improvement. Average maize grain yield per unit area in Ontario has increased at a rate of approximately 1.5 percent per year during the five-decade period since the introduction of hybrids in the 1940s, but average yields changed little during the five-decade period prior to hybrid introduction (see Figure 2.1). There is little doubt that hybrid vigor, which resulted in increased grain yield, decreased lodging, and increased stand uniformity of hybrids over open-pollinated varieties, contributed to the onset of the substantial and consistent yield improvements in maize. An average yield increase of 15 percent is commonly attributed to heterosis (Frey, 1971). It is not clear, however, whether heterosis per se has contributed beyond the initial increase.

In general, yield improvement can be attributed to genetic improvement, changes in cultural management, climatic change, and the interactions among these factors. Genetic improvement can be estimated from side-by-side comparisons of hybrids that are representative of the period under study. We estimated that the machine-harvestable grain yield of short-season maize hybrids grown in Ontario between 1959 and 1988 increased at a rate of 2.6 percent per year, and the hand-harvested grain yield, 1.7 percent per year (Tolle-

FIGURE 2.1. Average Maize Grain Yield in Ontario from 1892 to 1998

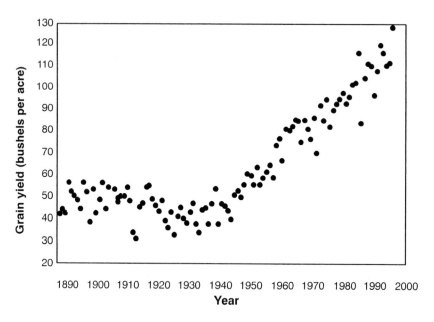

Source: Ontario Ministry of Agriculture and Food, Toronto, Ontario, Canada.

Note: One hundred bushels per acre = 5.3 Mg per ha at 0 percent grain moisture.

naar, 1989). Hence, genetic gain for this set of maize hybrids appears to be greater than the increase in the average provincial yield. We can only speculate why genetic gain may have been higher than the overall yield improvement. One possible reason for the greater genetic gain may be that the dominant area of maize production in Ontario has shifted from a longer to a shorter growing season, thereby reducing yield potential (Tollenaar and Dwyer, 1998). In contrast, the utilization of many improved cultural practices, including pest control, fertilization, and soil water management through underground drainage during the 1960s and 1970s (Daynard and Tollenaar, 1983), which made it possible to plant earlier and virtually extended the growing season, should have increased the yield harvested by commercial maize producers.

Increases in average maize yield of a magnitude similar to that recorded in Ontario after the introduction of hybrids have been reported for maize grown in the United States and Europe (Tollenaar, Dwyer, and McCullough, 1994). In contrast to Ontario, however, estimates of the contribution of genetic improvement to the overall yield improvement were in the range of 40 to 80 percent (e.g., Derieux et al., 1987; Duvick, 1992). Estimates of both yield improvement and the relative contribution of genetic gain to the overall yield improvement are influenced by genotype × environment (including crop management) interactions, and results are, in part, an artifact of the manner by which these interaction effects are explained. It is our contention that genotype × environment interaction effects on grain yield are the most prominent feature of yield improvement in maize. The most widely reported environmental factor that results in genotype × environment interactions, in comparisons of older and newer maize hybrids, is plant population density. For instance, Duvick (1984) showed that differences between older and newer genotypes, representing maize breeding from the 1930s to the 1970s, were small or nonexistent at low plant population densities, and that differences in grain yield between newer and older genotypes became greater as plant population density increased.

Maximum yields represent an indication of genetic yield potential of current maize hybrids. In the early 1980s, a farm-scale yield of 19.6 megagrams per hectare ($Mg \cdot ha^{-1}$) (0 percent grain moisture) was recorded by a maize producer in Illinois (Warsaw, 1985), and a yield of 15.5 $Mg \cdot ha^{-1}$ was recorded in large field research plots in Ontario (Stevenson, 1985). To the best of our knowledge, the 19.6 $Mg \cdot ha^{-1}$ yield is still the highest maize yield on record. Average yields of maize in the United States and Ontario in 1998 were less than 7.0 $Mg \cdot ha^{-1}$. Another example of the difference between potential and average maize yield is that of maize grown under low-stress growth room conditions. Mean yield over a four-year period of a short-season maize hybrid, released in 1976, was 11.5 $Mg \cdot ha^{-1}$, and daily incident photosynthetically active radiation was only 50 percent of that accumulated during summer months in Ontario (Tollenaar and Migus, 1984), indicating that the efficiency of maize production was three to four times higher under growth room than under field conditions. Overall, these results show that the potential yield of current maize hybrids is substantially greater than indicated by average commercial yield levels.

We will analyze the physiological basis of genetic improvement in maize, in general, and in short-season maize hybrids grown in Ontario, in particular. Yield improvement is the result of a more efficient capture and utilization of resources, and factors involved in the improved efficiency will be reviewed. The effect of stress tolerance and stand uniformity will also be discussed because improved efficiency in resource capture and utilization of newer hybrids is frequently apparent only under stressful conditions for maize growth. Finally, we will speculate about the potential for genetic improvement of maize in the future.

RESOURCE CAPTURE AND UTILIZATION

Resource Capture

Resource capture entails interception of incident solar radiation by the leaf canopy and uptake of nutrients and water by the roots. Interception of photosynthetic photon flux density (PPFD) is a function of leaf area index (LAI). Leaf area per plant has remained the same during the past six decades in the United States (Crosbie, 1982) and has increased only slightly from the late 1950s to the late 1980s in Ontario (Tollenaar, 1989), but the LAI of maize in commercial production has increased as the plant population density of commercially grown maize increased. However, PPFD interception of most maize hybrids grown at commercial population plant densities is greater than 90 percent when leaf area is fully expanded. The most important difference among maize genotypes in seasonal PPFD interception is the result of differences in the duration of the period when PPFD interception is greater than 90 percent. The duration of the period when PPFD interception is greater than 90 percent can be extended by either greater rates of area expansion during the period of leaf area expansion, delayed leaf senescence during the grain-filling period, or both. Nutrient and water uptake are related to the size and efficiency of the root system, as well as to the energy supply to the roots.

1. Total dry-matter accumulation and grain yield in maize, particularly in maize grown in short-season regions, could be increased substantially by advancing the date at which canopy PPFD interception is greater than 90 percent (Tollenaar, 1983). Enhanced seedling vigor could advance time of canopy closure and increase seasonal PPFD interception. Our results with short-season maize hybrids have shown that total aboveground dry matter before the twelve-leaf stage of a newer hybrid was lower than that of an older hybrid in growth cabinet studies (Tollenaar, Mihajlovic, and Aguilera, 1991; McCullough et al., 1994), and aboveground dry matter of newer hybrids was also lower than that of older hybrids in a field study that involved three older and three newer hybrids (Tollenaar, unpublished data). Results of the growth cabinet studies indicated that early vigor was inversely associated with the root:shoot ratio. Less dry matter is allocated to the leaves if a greater proportion of dry matter is allocated to the roots, which results in a smaller LAI and lower PPFD interception because PPFD interception is proportional to LAI during early phases of development, when LAI is small. Hence, a genotype with a high root:shoot ratio would be expected to have a relatively low early vigor if leaf photosynthetic rate does not differ among genotypes. Leaf photosynthetic rate did not differ between the two hybrids (Tollenaar, Mihajlovic, and Aguilera, 1991; McCullough et al., 1994), and plant dry-matter accumulation and rate of leaf area expansion during early development were inversely related to the root:shoot ratio.

2. The effect of extending the period at which PPFD interception is greater than 90 percent by delaying leaf senescence during the grain-filling period is, at least theoretically, smaller than that of advancing the date of canopy closure because daily solar irradiance is less during later phases of the grain-filling period than during the period of leaf area expansion, which is close to the summer solstice (Tollenaar, 1983). In contrast to the effect of early vigor on yield differences among older and newer maize genotypes, however, delayed leaf senescence, or "stay green," is associated with yield improvement of maize hybrids in North America (e.g., Crosbie, 1982; Tollenaar, 1991; Duvick, 1997). Most of the differences in

dry-matter accumulation between older and newer hybrids can be attributed to differences during the grain-filling period (Tollenaar, 1991; Tollenaar and Aguilera, 1992). The stay-green characteristic of a newer relative to an older hybrid has been associated with an improved source:sink ratio during the grain-filling period (Rajcan and Tollenaar, 1999a), which has led to both higher total dry-matter accumulation and lower stem lodging (Tollenaar, 1989).

3. Nutrient uptake is related to size and efficiency of the root mass and energy supply. Few data on root mass during later phases of maize development have been reported in the literature. A comparison of the older hybrid Pride 5 with the newer hybrid Pioneer 3902, grown in a hydroponic system in the field (Tollenaar and Migus, 1984) in 1992 and 1993, showed that the root:shoot ratio was approximately 20 percent greater in the newer than in the older hybrid during the grain-filling period (Nissanka, 1995). A higher source:sink ratio and higher bottom-stem weight at maturity in the newer than in the older hybrid (Rajcan and Tollenaar, 1999a) may indicate that assimilate supply to the roots was greater in the newer hybrid. Rajcan and Tollenaar (1999b) reported that the proportion of grain nitrogen (N) derived from postsilking N uptake was 60 percent for the newer hybrid and 40 percent for the older hybrid. Continuous N uptake during the grain-filling period has been associated with the ability to maintain root growth after silking (Mackay and Barber, 1986), which may be a function of assimilate supply. Whereas we found differences in nitrogen use efficiency between an older and a newer hybrid, Castleberry and colleagues (1984) reported that relative differences among maize hybrids from the 1930s to the 1980s were similar when grown under high and low soil nutrient conditions, and Duvick (1984) reported no hybrid $\times$ nitrogen interactions for hybrids representing five decades of yield improvement in the United States when grown at two nitrogen rates.

4. Water uptake also may be related to root mass. Dwyer, Stewart, and Tollenaar (1992) showed that when the older hybrid Pride 5 and the newer hybrid Pioneer 3902 were grown in the field under moisture stress, leaf photosynthetic rates were higher in the newer than in the older hybrid until 11:00 a.m., but not after 11:00. Nissanka and colleagues (1997) showed, for plants grown in a canopy enclosure, that the decline in plant photosynthesis during a drying cycle around

silking occurred, on average, one day earlier in the older than in the newer hybrid, and that the recovery of canopy photosynthesis after rewatering was greater in the newer than in the older hybrid. Results of the latter study may indicate that differences between the two hybrids are related to the ability to tolerate low water potential, as canopy photosynthesis of the newer hybrid continued at much lower stem water potentials than that of the older hybrid. This contention is corroborated by unpublished results of John S. Sperry (University of Utah) that showed a much lower loss in percent stem conductivity for Pioneer 3902 than for Pride 5 when xylem pressure was in the range of -0.5 to -2.5 megapascals (MPa).

Resource Utilization

Four potential avenues for increasing resource utilization are (1) an increase in gross leaf photosynthetic rate, (2) an increase in gross canopy photosynthetic rate due to a more uniform distribution of incident solar radiation across the leaf canopy, (3) a reduction in plant respiration, and (4) an increase in the proportion of total above-ground dry matter that is allocated to the grain at maturity (i.e., harvest index). Information on differences between older and newer maize hybrids in respect to gross leaf photosynthesis and plant respi-ration is predominantly derived from results of studies with the series of older and newer Ontario hybrids (cf. Tollenaar, 1989).

1. An increase in gross leaf photosynthesis may be the result of either an increase in photosynthesis at saturating PPFD (i.e., P_{max}), an increase of photosynthesis at low PPFD (i.e., quantum efficiency), or both. Simulations of the effects of increased leaf photosynthesis on maize growth and yield indicate that an increase in P_{max} has a relatively small effect on yield (Boote and Tollenaar, 1994). Under optimal conditions, differences in leaf net photosynthesis at high PPFD between old and new hybrids were not significant (Crosbie, 1982; Dwyer and Tollenaar, 1989). The day after a cold night (i.e., a minimum temperature of $9.6°C$), however, net leaf photosynthesis at high PPFD was reduced more in older than in newer hybrids (Dwyer and Tollenaar, 1989). The effect of a change in quantum efficiency on grain yield of maize is much greater than that of a change in P_{max} (Boote and Tollenaar, 1994). We have not observed any differences among older and newer hybrids in respect to either quantum efficiency

or carbon dioxide (CO_2) fixation per electron transported in the thylakoid membrane (Earl and Tollenaar, 1998b).

2. A more uniform distribution of incident solar radiation across the crop canopy can result in an increase in gross canopy photosynthetic rate (Tollenaar and Dwyer, 1998). An increase in leaf angle results in a more uniform distribution of solar radiation across the canopy, and an increase in leaf angle have been reported for Corn Belt hybrids between 1930 and 1990 (Crosbie, 1982; Duvick, 1997). Effect of leaf angle on canopy photosynthesis can be estimated from the change in PPFD distribution across the leaf canopy and the photosynthesis-PPFD response curve: our estimates indicate a yield improvement in the order of 20 percent when leaf angle increases from 30 to 60 degrees (Tollenaar and Dwyer, 1998). In contrast to maize hybrids in the U.S. Corn Belt, canopy architecture of short-season hybrids in central Ontario had not changed by the late 1980s (Tollenaar and Aguilera, 1992), although leaf angle appears to have increased in hybrids introduced during the 1990s (Tollenaar, unpublished data).

3. The two major components of respiration are growth respiration, which is a function of the composition of the plant dry matter, and maintenance respiration, which is a function of the energy required to maintain plant function and structure (Penning de Vries, Brunsting, and van Laar, 1974; Penning de Vries, 1975). Little has been reported on changes in plant composition associated with yield improvement in maize. Vyn and Tollenaar (1998) reported that changes in grain composition associated with yield improvement in Ontario were minor, and Duvick (1997) reported that grain protein concentration of U.S. Corn Belt hybrids declined by 1.8 percent per year from the 1930s to the 1990s. A decline in protein concentration will result in a reduction of growth respiration per unit grain, but no data have been reported on total plant respiration and its components. The overall effect of a small change in composition, however, will not have a large impact on grain yield. For instance, "high oil corn" (HOC) has been recently introduced in the United States and Canada. The grain of HOC maize hybrids has nearly twice the oil concentration and substantially higher protein than conventional maize hybrids (Butzen and Cummings, 1999). Based on principles outlined by Penning de Vries and colleagues (1974), it can be estimated that an

increase in grain oil concentration from 3 to 8 percent will result in only a 6 percent decline in yield. Even less has been reported on changes in maintenance respiration between older and newer maize genotypes. Earl and Tollenaar (1998a) reported a strong negative correlation between mature-leaf respiration during the growing season and total season dry-matter accumulation among three old and three new Ontario maize hybrids. Nissanka and colleagues (1997) reported that the ratio of aboveground canopy respiration and gross canopy photosynthesis around silking was 28 percent greater in the older hybrid Pride 5 than in the newer hybrid Pioneer 3902 when grown under relatively optimal conditions. As the composition of the vegetative dry matter of these hybrids was similar (Rajcan and Tollenaar, 1999b), the apparent difference in respiration between the two hybrids may be attributed to maintenance respiration.

4. Differences in the response of grain yield to plant density among maize hybrids representing various eras of breeding have led some to conclude that harvest index has been positively associated with genetic improvement in maize (Russell, 1985). Harvest index declines when plant density is increased beyond the optimum plant density for grain yield in maize, and the optimum plant density for grain yield is lower for old than for new hybrids, which may explain the association between harvest index and era of release for hybrids grown at high plant density (Tollenaar, Dwyer, and McCullough, 1994). Harvest index did not differ among hybrids representing three decades of yield improvement in Ontario when yields were compared at the optimum plant population density for each of the hybrids (Tollenaar, 1989).

STRESS TOLERANCE AND STAND UNIFORMITY

Stress Tolerance

We define stress as a factor that causes, through either its presence or its absence, a reduction in plant grain yield. Consequently, stress is defined in terms of the plant's response to a causal factor. Stress can be alleviated either through management practices that influence the causal factor or by modifying the plant such that the impact of the causal factor on plant grain yield is reduced. Common stresses that can influence plant grain yield are low soil moisture, low soil N, low absorbed PPFD, and low or high air temperatures. An increase in either the number of maize plants per unit area (i.e., plant population density) or number and size of weeds within a maize stand will enhance the competition between plants for a variety of resources within the maize canopy, and, consequently, stresses of this nature are termed compound stresses. Grain yield per unit area of a maize canopy shows a curvilinear response to plant density, with a maximum yield at the optimum plant density. Grain yield per unit area increases with increasing plant density until the increase in yield attributable to the addition of plants is not greater than the decline in mean yield per plant. Consequently, high yield per unit area and plant stress are inevitably linked. When the stress is very high, a binomial distribution of grain yield per plant becomes apparent, with an increasing proportion of plants that have no grain-bearing ear. Under extremely high stress conditions, effects of stress on yield may be quite different from those under commercial conditions for maize production. The former conditions we call survival environments, and the latter we call production environments, and, arbitrarily, we set the yield dividing the two environments at 2 to 3 $Mg \cdot ha^{-1}$. Our discussion on the effects of stress on grain yield will be limited to maize grown in production environments.

Results of our studies with short-season hybrids representing grain yield improvement in central Ontario have shown that newer hybrids are more tolerant to stress than older hybrids. The differential response to stress between older and newer hybrids has been shown for low night temperature during the grain-filling period (Dwyer and Tollenaar, 1989), for low soil moisture in the field (Dwyer, Stewart, and Tollenaar, 1992) and under controlled-environment conditions (Nissanka, Dixon, and Tollenaar, 1997), low soil-N (McCullough et al., 1994), and for herbicide (Tollenaar and Mihajlovic, 1991), in addition to compound stresses such as plant population density (Tollenaar, 1992) and weed interference (Tollenaar, Aguilera, and Nissanka, 1997).

Stand Uniformity

Although stand uniformity of field crops has long been recognized as an important aspect of high-yielding cultivars, and breeders have been pursuing it since the beginning of modern agriculture, it also has been frequently claimed that cultivar mixtures can give higher and more stable yields than pure lines in cereal crops (Wolfe, 1985). Commercial maize genotypes in the U.S. Corn Belt evolved from open-pollinated varieties before the 1930s, to double-cross hybrids from the 1930s to the 1960s, and to single-cross hybrids since the late 1960s. The evolution of open-pollinated varieties of the 1930s to the single-cross hybrids of the 1990s was associated with increases in both rate of yield improvement and area of adaptation (Troyer, 1996). This evolution was also associated with a decrease in genetic variability within and among commercial maize genotypes: hundreds of open-pollinated varieties were grown until the 1930s in the U.S. Corn Belt, but the parentage of virtually all relatively recent commercial U.S. hybrids involves only six inbred lines or their close relatives (Goodman, 1990). We have found that increases in yield and stress tolerance of more recent maize hybrids grown in Ontario are associated with enhanced crop stand uniformity (Wu, 1998). The increase in area of adaptation and the apparent increase in stress tolerance of newer commercial maize hybrids, in comparison with older hybrids and open-pollinated populations, contradict the notion that genotypic and phenotypic variability are positively associated with yield stability.

Results of our studies indicate that stand uniformity and stress tolerance are highly associated (Wu, 1998). The relationship between stand uniformity and stress tolerance was investigated in an older and a more recent hybrid grown at 3.5 and 11 plants per square meter (plants/m^2). At each plant population density, seeds were sown either all on the same day to produce uniform stands (control) or on three dates spaced one thermal leaf unit (Tollenaar, Daynard, and Hunter, 1979) apart to produce nonuniform stands (treatment); the control and all border plants were seeded at the second seeding date of the nonuniform stand. At physiological maturity, a significant ($P < 0.05$) (plant population density $\times$ treatment) interaction was observed for grain yield, total aboveground dry matter, and CV of the individual plant yield and dry matter. At the low plant population density, stands of the uniform and nonuniform sowing did not differ for any of the measured traits. At the high plant population density, nonuniform stands yielded less than uniform stands in both hybrids, and the yield difference between the newer and older hybrid was 30 percent in the uniform stand and 46 percent in the nonuniform stand. In a second experiment, the association between genetic improvement and stand uniformity among six maize hybrids released from 1959 to 1995 was explored. Hybrids were grown at 3.5, 7.0, and 11 plants/m^2 and two soil N levels. A high and negative correlation (R^2 = 0.73) was found between plant-to-plant variability for individual plant grain yield and total grain yield per unit area at the medium and high plant densities. Results of both studies show that density-tolerant hybrids exhibit less plant-to-plant variability than nontolerant ones at plant densities close to or higher than their optimum densities.

CONCLUSIONS

A retrospective analysis can indicate factors that have contributed to the genetic yield improvement, but not all of those factors will contribute to future yield improvements. For instance, first, the initial yield improvement of maize hybrids over open-pollinated varieties was attributable, in part, to heterosis (i.e., an average yield increase of 15 percent is commonly attributed to heterosis [Frey, 1971]), but it is not clear whether heterosis has contributed beyond the initial increase and whether it will contribute in the future. Second, the incorporation of a defensive trait (e.g., pest resistance) is a one-time contribution to yield, and a continuous breeding effort may be required to maintain yield potential at the same level in order to stay one step ahead of the pest organism as it circumvents the plant's resistance. We have not observed large differences in pest resistance among genotypes in our studies of older and newer hybrids in Ontario (Tollenaar, Dwyer, and McCullough, 1994), but several single-gene traits that have recently been introduced in maize through biotechnology fall in this category. Third, in a similar vein, improvement in physiological traits such as "stay green" and a more uniform distribution of incident PPFD across the crop canopy are largely one-time improvements.

In contrast, some of the factors that have shown no or little improvement in the past could contribute to future yield improvement: (1) Increased chilling tolerance should be a high priority in maize breeding programs, in particular, for relatively cool, short-season environments. The reduction in leaf photosynthesis following a cool, but not cold, night during the grain-filling period (Dwyer and Tollenaar, 1989) indicates that increased chilling tolerance may be able to increase seasonal dry-matter accumulation substantially, as cool nights are common during the last two months of the growing season in Ontario (i.e., from middle of August to middle of October). (2) High early plant vigor will increase PPFD interception during early phases of development, but the increased vigor should be the result of higher plant photosynthesis and should not be due to a lower root:shoot ratio. Increased early vigor may result from increased chilling tolerance as described under (1). (3) If the genetic improvement in grain yield of maize continues, the rate of improvement in grain yield could be proportionally greater than the improvement in seasonal dry-matter accumulation (i.e., harvest index will increase). Seasonal canopy photosynthesis can increase without a change in plant stature or plant leaf area, and the structural framework supporting the enlarged grain factory (i.e., stems and roots) can support the higher grain production with a less than proportional increase in dry matter. It should be noted, however, that harvest indices much higher than 50 percent have not yet been documented under any circumstance for any crop.

Increased stress tolerance, combined with increased stand uniformity under stress conditions, will likely continue to provide the highest potential for yield improvement in maize in the next decades. The record yields for maize were achieved more than fifteen years ago, as was noted earlier, and closing the gap between this measure of yield potential and average yields by commercial maize producers probably offers the greatest opportunity for future yield improvement.

REFERENCES

Boote, K.J. and Tollenaar, M. (1994). Modeling genetic yield improvement. In Boote, K.J., Bennett, J.M., Sinclair, T.R., and Paulsen, G.M. (Eds.), *Physiology and determination of crop yield* (pp. 533-565). Madison, WI: ASA-CSSA-SSSA.

Butzen, S. and Cummings, M. (1999). Corn grain protein. *Crop Insights* 9(11): 1-5.

Castleberry, R.M., Crum, C.W., and Krull, C.F. (1984). Genetic yield improvement of U.S. maize cultivars under varying fertility and climatic environments. *Crop Science* 24: 33-36.

Crosbie, T.M. (1982). Changes in physiological traits associated with long-term breeding efforts to improve grain yield of maize. In Loden, H.D. and Wilkinson, D. (Eds.), *Proceedings 37th annual corn and sorghum industry research conference* (pp. 206-233). Washington, DC: ASTA.

Daynard, T.B. and Tollenaar, M. (1983). Prospects for improving the productivity of early maturing maize. In Gallais, A. (Ed.), *Physiolgie du maïs* (pp. 535-570). Paris, France: INRA.

Derieux, M., Darrigrand, M., Gallais, A., Barriere, Y., Bloc, D., and Montalant, Y. (1987). Estimation du progres genetique realise chez le maïs grain en France entre 1950 et 1985. *Agronomie* 7: 1-11.

Duvick, D.N. (1984). Genetic contributions to yield gains of U.S. hybrid maize, 1930 to 1980. In Fehr, W.R. (Ed.), *Genetic contributions to yield gains of five major crop plants* (pp. 1-47). Madison, WI: CSSA-ASA.

Duvick, D.N. (1992) Genetic contributions to advances in yield of U.S. maize. *Maydica* 37: 69-79.

Duvick, D.N. (1997). What is yield? In Edmeades, G.O., Banziger, M., Mickelson, H.R., and Peña-Valdivia, C.B. (Eds.), *Developing drought- and low-N-tolerant maize* (pp. 332-335). Mexico DF, Mexico: CIMMYT.

Dwyer, L.M., Stewart, D.W., and Tollenaar, M. (1992). Analysis of maize leaf photosynthesis under drought stress. *Canadian Journal of Plant Science* 72: 477-481.

Dwyer, L.M. and Tollenaar, M. (1989). Genetic improvement in photosynthetic response of hybrid maize cultivars, 1959 to 1988. *Canadian Journal of Plant Science* 69: 81-91.

Earl, H.J. and Tollenaar, M. (1998a). Differences in rates of mature leaf respiration among commercial maize hybrids. *Field Crops Research* 59: 9-19.

Earl, H.J. and Tollenaar, M. (1998b). Relationship between thylakoid electron transport rate and photosynthetic CO_2 uptake in leaves of three maize (*Zea mays* L.) hybrids. *Photosynthesis Research* 58: 245-257.

Frey, K.J. (1971). Improving crop yields through plant breeding. In Eastin, J.D. and Munson, R.D. (Eds.), *Moving of the yield plateau,* Special publication number 20 (pp. 15-58). Madison, WI: ASA-CSSA-SSSA.

Goodman, M.M. (1990). Genetic and germplasm stocks worth conserving. *Journal of Heredity* 81: 11-16.

Mackay, A.D. and Barber, S.A. (1986). Effect of nitrogen on root growth of two genotypes in the field. *Agronomy Journal* 78: 699-703.

McCullough, D.E., Girardin, P., Mihajlovic, M., Aguilera, A., and Tollenaar, M. (1994). Influence of N supply on development and dry matter accumulation of an old and new maize hybrid. *Canadian Journal of Plant Science* 74: 471-477.

Nissanka, S.P. (1995). The response of an old and a new maize hybrid to nitrogen, weed and moisture stress. PhD thesis, University of Guelph, Canada.

Nissanka, S.P., Dixon, M.A., and Tollenaar, M. (1997). Canopy gas exchange response to moisture stress in old and new maize hybrids. *Crop Science* 37: 172-181.

Penning de Vries, F.W.T. (1975). The cost of maintenance processes in plant cells. *Annals of Botany* (London) 39: 77-92.

Penning de Vries, F.W.T., Brunsting, A.H.M., and van Laar, H.H. (1974). Products, requirements, and efficiency of biosynthesis: A quantitative approach. *Journal of Theoretical Biology* 45: 339-377.

Rajcan, I. and Tollenaar, M. (1999a). Source-sink ratio and leaf senescence in maize. I. Dry matter accumulation and partitioning during the grain-filling period. *Field Crops Research* 60: 245-253.

Rajcan, I. and Tollenaar, M. (1999b). Source-sink ratio and leaf senescence in maize. II. Metabolism of nitrogen and soluble carbohydrates during the grain-filling period. *Field Crops Research* 60: 255-265.

Russell, W.A. (1985). Evaluations for plant, ear, and grain traits of maize cultivars representing different eras of breeding. *Maydica* 30: 85-96.

Stevenson, C.K. (1985). Maximum yield corn research. In Munson, R.D. (Ed.), *Physiology, biochemistry, and chemistry associated with maximum yield corn* (pp. 190-191). Workshop proceedings, Foundation for Agronomic Research and Potash Phosphate Institute. St. Louis, MO, November 11-12.

Tollenaar, M. (1983). Potential vegetative productivity in Canada. *Canadian Journal of Plant Science* 63: 1-10.

Tollenaar, M. (1989). Genetic improvement in grain yield of commercial maize hybrids grown in Ontario from 1959 to 1988. *Crop Science* 29: 1365-1371.

Tollenaar, M. (1991). The physiological basis of the genetic improvement of maize hybrids in Ontario from 1959 to 1988. *Crop Science* 31: 119-124.

Tollenaar, M. (1992). Is low plant density a stress in maize? *Maydica* 37: 305-311.

Tollenaar, M., and Aguilera, A. (1992). Radiation use efficiency of an old and a new maize hybrid. *Agronomy Journal* 84: 536-541.

Tollenaar, M., Aguilera, A., and Nissanka, S.P. (1997). Grain yield is reduced more by weed interference in an old than in a new maize hybrid. *Agronomy Journal* 89: 239-246.

Tollenaar, M., Daynard, T.B., and Hunter, R.B. (1979). The effect of temperature on rate of leaf appearance and flowering date in maize. *Crop Science* 19: 363-366.

Tollenaar, M. and Dwyer, L.M. (1998). Physiology of maize. In Smith, D.L. and Hamel, C. (Eds.), *Crop yield, physiology and processes* (pp. 169-204). New York: Springer.

Tollenaar, M., Dwyer, L.M., and McCullough, D.E. (1994). Physiological basis of genetic improvement of corn. In Slafer, G.A. (Ed.), *Genetic improvement of field crops* (pp. 183-236). New York: Marcel Dekker.

Tollenaar, M. and Migus, W. (1984). Dry matter accumulation of maize grown hydroponically under controlled-environment and field conditions. *Canadian Journal of Plant Science* 64: 475-485.

Tollenaar, M. and Mihajlovic, M. (1991). Bromoxynil tolerance during the seedling phase is associated with genetic yield improvement in maize. *Canadian Journal of Plant Science* 71: 1021-1027.

Tollenaar, M., Mihajlovic, M., and Aguilera, A. (1991). Temperature response of dry matter accumulation, leaf photosynthesis, and chlorophyll fluorescence in an old and a new maize hybrid during early development. *Canadian Journal of Plant Science* 71: 353-359.

Troyer, A. F. (1996). Breeding widely adapted, popular maize hybrids. *Euphytica* 92: 163-174.

Vyn, T.J. and Tollenaar, M. (1998). Changes in chemical and physical quality parameters of maize grain during three decades of yield improvement. *Field Crops Research* 59: 135-140.

Warsaw, H. (1985). High yield corn production. In Munson, R.D. (Ed.), *Physiology, biochemistry, and chemistry associated with maximum yield corn* (pp. 193-199). Workshop proceedings, Foundation for Agronomic Research and Potash Phosphate Institute, St. Louis, MO, November 11-12.

Wolfe, M.S. (1985). The current status and prospects of multiline cultivars and variety mixtures for disease resistance. *Annual Review of Phytopathology* 23: 251-273.

Wu, J. (1998). On the relationship between plant-to-plant variability and stress tolerance in maize (*Zea mays* L.) hybrids from different breeding eras. MSc thesis, University of Guelph, Canada.

Chapter 3

Processes Affecting Maize Grain Yield Potential in Temperate Conditions

James R. Kiniry
María E. Otegui

INTRODUCTION

Crop modelers and crop breeders have a common interest in identifying processes and traits that control grain yield. Modelers need to derive equations to predict yields of cultivars with different climatic conditions on several soil types. Breeders need to know the yield-limiting processes/traits to most effectively and efficiently select for increased yield. The objectives of this chapter are to describe a mechanism for characterizing maize (*Zea mays* L.) seed number production and to discuss the importance of individual seed weight and nonstructural carbohydrate storage for grain yield. Such information should be helpful in selecting for increased grain yields at different sites.

SETTING NUMBER OF SEEDS PER EAR

During the period from ear initiation to silking, the number of ovules is determined. However, ovule number may not be important for final grain yield. For maize grown under plant populations typically used in temperate humid regions ($\geq$ 5.5 plants/square meter [m^2]), this potential number appears always to exceed the actual final number of filled seeds per ear (Otegui, 1995). Whereas Bonhomme and colleagues (1984) found that the number of ovules increased with increasing time to silking, we have found no significant relationship between the number of ovules and the final number of filled seeds (Otegui, 1997; Otegui and Melón, 1997). Nevertheless, this trend cannot be taken as a recommendation to reduce the number of florets per ear because this trait may eventually limit seed set per plant under favorable environmental conditions (Lafitte and Edmeades, 1995) and low plant populations (Ruget, 1993; Tollenaar, Dwyer, and Stewart, 1992; Andrade et al., 1999).

Much of the early work on maize seed number involved looking at the synchrony between pollen shedding and silk emergence. DuPlessis and Dijkhuis (1967) showed a linear decline in seeds per plant as silking was delayed from six to seventeen days after pollen shedding. Hall and colleagues (1982) showed a strong association between pollen availability and the number of seeds. Similarly, Edmeades and co-workers (1993) found that the anthesis-silking interval (ASI) was a valuable indicator of stress tolerance in maize selections. However, even when water and nutrients were not limiting growth, seed set per plant was generally smaller than the number of fertilized ovaries (Otegui, 1995; Otegui, Andrade, and Suero, 1995). Most commercial hybrids have seeds on the tips of the ears that were fertilized but did not develop (i.e., have hard endosperm but are small). Even under severe drought, which impaired pollination of many silks due to protandry (Hall et al., 1982; Bolaños and Edmeades, 1993) and resulted in very low grain yields, maize ears had more fertilized ovules than they could fill. Pollen shedding and fertilization did not limit grain yield (Westgate and Boyer, 1986).

The approximate duration of the critical period for seed set was determined with shading stress (Kiniry and Ritchie, 1985) and drought stress (Grant et al., 1989) experiments. These studies defined a period immediately following silking as the most important for seed set. Otegui and Bonhomme (1998) determined the actual thermal time requirement for ear growth of many hybrids grown in different environments. In temperate conditions, the duration in days of the period for ear growth was very stable, even for a wide range of sowing dates (August 20 to November 20). During this period, the presilking environment (silking −227 °Cday*) determines the number of seed-bearing ears per plant (prolificacy). Abortion of fertilized ovaries takes place mostly during the first one or two weeks after silking (silking + 100 °Cday) but may continue up to the end of the lag phase of grain filling (Cirilo and Andrade, 1994). It is during this postfertilization seed development phase that final seed number is determined (Hall, Lemcoff, and Trápani, 1981).

Sucrose supply to the ear during this interval is critical for seed set. Boyle, Boyer, and Morgan (1991) found that infusing sucrose solution into stems of plants suffering drought stress during the five to seven days following silking prevented a decrease in seed number. Additional stem infusion studies supported this finding (Schussler and Westgate, 1995; Zinselmeier, Lauer, and Boyer, 1995). Likewise, it is the carbohydrate flux during the critical period that controls seed number, independent of the quantity of stored carbohydrate prior to anthesis. Increasing the stored carbohydrate at anthesis by growing plants at low population density did not decrease seed abortion in response to drought (Schussler and Westgate, 1994). Plants fill only the number of seeds that can be supported by carbohydrate from concurrent photosynthesis during this period.

The number of seeds set should be linearly related to the assimilate flux into the ear during the critical period, if we assume a constant amount of assimilate is required to support each developing fertilized ovule. Assuming there is a stable amount of carbohydrate produced per unit intercepted light (radiation use efficiency), this would support findings of a linear response of seed number per plant (SNP) to photosynthetically active radiation intercepted per plant (IPARP) during the critical period. Such a linear response has been shown for sorghum [*Sorghum bicolor* (L.) Moench], by Vanderlip and co-workers (1984), and for wheat (*Triticum aestivum*), by Fischer (1985). For maize cropped in the temperate region of Temple, Texas (31°06′N, 97°20′W) ,and the cool temperate region near College Park, Pennsylvania (40°48′N, 77°52′W), Kiniry and Knievel (1995) established a general model (see equation 3.1) for seed set

---

*Degree days throughout this chapter have a base temperature of 8 °C.

prediction based on the amount of IPARP during the ten-day period after silking (see Figure 3.1):

$$SNP = 165 + 253 \text{ IPARP}; \quad r^2 = 0.77 \ (n = 72) \qquad (3.1)$$

Similar equations could be derived from data obtained in a plant population experiment by Andrade, Uhart, and Frugone (1993) for the cool temperate region of Balcarce, Argentina (37°45′S, 58°18′W) (equation 3.2):

$$SNP = 115 + 257 \text{ IPARP}; \quad r^2 = 0.83 \ (n = 18) \qquad (3.2)$$

and by Otegui and Bonhomme (1998) for the nonprolific hybrids cropped in temperate (32° to 37°S, 58° to 62°W, Argentina) and high-latitude (48°51′N, 1°58′E, France) environments (equations 3.3 and 3.4):

$$SNP = 184 + 278 \text{ IPARP}; \quad r^2 = 0.29 \ (n = 18) \quad \text{Argentina} \qquad (3.3)$$
$$SNP = 41 + 498 \text{ IPARP}; \quad r^2 = 0.55 \ (n = 17) \quad \text{France} \qquad (3.4)$$

Differences between models set for Texas and Argentina are within the range of responses measured for different hybrids by Kiniry and Knievel (1995). On the other hand, models from these temperate regions differed significantly ($P < 0.05$) from that obtained for a high-latitude environment in France (see Figure 3.1). In France, plants were growing always under less than 1 megajoule per plant (MJ/plant) per day. These extreme conditions are seldom met in temperate environments (see Figure 3.1). At intermediate latitudes (30° to 40°), low IPARP values are reached only when high plant populations (plants/m^2 ≥ 9) are used to maximize grain yields (see Figure 3.2). Then, small reductions in IPARP promote a nearly vertical drop in the number of seeds per plant (Andrade, Otegui, and Vega, 2000; Andrade, Cirilo, and Echarte, 2000), indicating that maize requires a threshold carbohydrate flux into the ear to produce any seeds.

Genotypic differences also exist for the response of SNP to IPARP (Kiniry and Knievel, 1995). Cultivars obviously have a limit to the number of seeds per plant that they can produce (Otegui and Melón, 1997), even at very low planting densities. The line for seed number as a function of intercepted light has a plateau at high IPARP. The

FIGURE 3.1. Relationship Between Seed Number per Plant (SNP) and Intercepted Photosynthetically Active Radiation per Plant (IPARP) During the Critical Period for Kernel Set

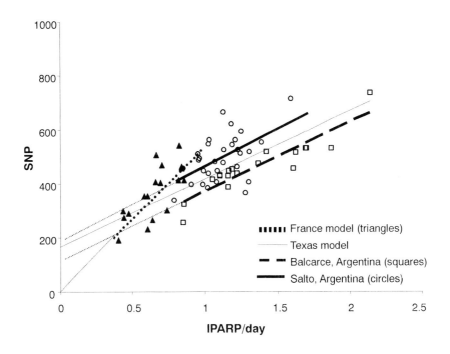

Source: Adapted from Kiniry and Knievel (1995) and Otegui and Bonhomme (1998).

value for this plateau, as well as for the break point at low IPARP, is most likely to be genotype dependent. A large y-intercept may indicate improved tolerance to increased plant population.

While Kiniry and Knievel (1995) found that several hybrids showed a common response line for SNP as a function of IPARP, other hybrids showed different lines. In this context, the flex-eared hybrid used in their work did not show the expected response: the slope of the relationship between SNP and IPARP was not steeper than the general line fitted for several hybrids. Knowing this response can be useful when selecting for increased yields and when optimizing population density.

FIGURE 3.2. Response of Seed Number per Plant (SNP, ◈) and Grain Yield (•) to Plant Growth Rate During the Critical Period for Kernel Set

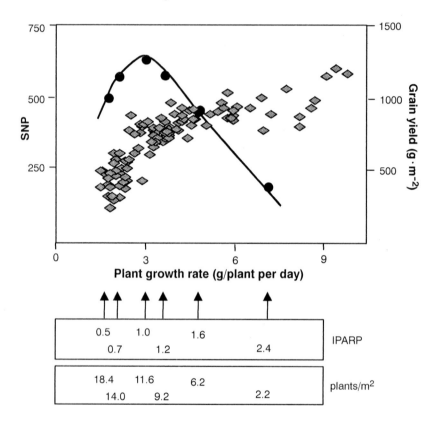

Source: Adapted from Andrade et al. (1996, p. 237) and Andrade et al. (1999, p. 456).

Note: The amount of light intercepted by the plants (IPARP, in MJ/plant per day) and plant density (plants/m²) are indicated for different plant growth rates.

In the absence of drought or nutrient limitations, such as under high management conditions (e.g., irrigation), increased plant density increases the number of seeds per unit area. The high population densities used for irrigated maize in temperate regions support these results.

Under dryland situations, this trend must be balanced with the greater probability of drought stress near silking with higher leaf area index. Thus, optimum dryland planting densities are lower and should decrease with increasing probability of drought. Such trends are commonplace and have resulted from years of experience by producers.

IMPORTANCE OF STORED CARBOHYDRATE FOR YIELD PRODUCTION WITH LATE-SEASON STRESS

If severe drought occurs after anthesis, stored assimilate may be the only carbohydrate available for grain growth of maize. Under these conditions, the quantity of stored carbohydrate could be a major constraint to grain growth. Translocation to grain continues even when the drought is sufficiently severe to halt photosynthesis (McPherson and Boyer, 1977).

Maize shows less variation in nonstructural carbohydrate (NCO) in the stem at anthesis than sorghum [*Sorghum bicolor* (L.) Moench]. McBee and colleagues (1983) found sorghum cultivars with NCO in the culm ranging from 60 to 350 grams per kilogram $(g \cdot kg^{-1})$ at fifteen days postanthesis. For maize, decreases in stem soluble solids following silking ranged from 200 to 310 $g \cdot kg^{-1}$ of stem dry weight at silking (Daynard, Tanner, and Hume, 1969; Campbell and Hume, 1970). Reported concentrations of NCO in maize stems near silking range from 150 to 250 $g \cdot kg^{-1}$ (Williams et al., 1968; Campbell and Hume, 1970; Swank et al., 1982; Jones and Simmons, 1983).

Nonstructural carbohydrates disappearing from vegetative organs are used for synthesis of structural material, for plant maintenance, and for growth of the grain and ear. The amount of NCO used for grain growth and the efficiency of production of grain per unit NCO used are critical factors in understanding grain yield production under stress conditions. Beginning with glucose and amino acids, the theoretical conversion efficiency of glucose into reproductive growth, including the maintenance requirement, is 0.65 g of ear per g of glucose for maize (Penning de Vries, van Laar, and Chardon, 1983). If maintenance costs are removed, this efficiency is 0.73.

Over two years, we imposed severe shading stress on maize beginning seven to eight days after anthesis (Kiniry et al., 1992). Treatments included shading and control, both with plants intact and with plants having their ears removed at the start of the shading treatment. This allowed the determination of the amount of carbohydrate change in the vegetative plant, without an actively growing reproductive sink. The decrease in NCO in such plants that were shaded was taken as the costs of respiration of the vegetative portions of the plants. Destructive sampling was done and dry-weight changes in stems, leaves, and ears were measured. Nonstructural carbohydrates were also measured. Stem dry weights increased in dry weight as well as in NCO during the week following anthesis (see Figure 3.3). During this interval, stem dry weight increased 25 percent in 1988 and 22 percent in 1989. In the same interval, NCO concentration increased from 24 to 41 percent in 1988 and from 19 to 23 percent in 1989. The decrease in stem weight of shaded plants with ears in 1988 was 28.8 g/plant during the shading treatment. Since the value for shaded plants without ears was 12.9 g/plant, it was estimated that 12.9 g/plant was used for maintenance and 15.9 g/plant for ear growth. Comparing estimated stem dry-weight decrease from seven to twenty-nine days with the estimated decrease in NCO, both for shaded plants with ears, the first decrease was $390 \text{ g} \cdot \text{kg}^{-1}$ of stem dry weight, whereas the latter decrease was $370 \text{ g} \cdot \text{kg}^{-1}$ of the stem dry weight. This close agreement implies that the reduction in stem dry weight was essentially a decrease in NCO.

The efficiency of grain production from dry weight lost from stem and leaf laminae in 1988, deleting estimated dry weight for maintaining the vegetative organs, was 0.51 g of grain per g of dry weight. When maintenance costs were included, this efficiency was 0.26.

It appeared that nonstructural carbohydrate stored prior to grain filling had little buffering capacity for grain growth when severe shading occurred after anthesis. Much of the stored assimilate was consumed for maintenance. Grain production from this stored assimilate was less efficient than predicted from theoretical calculations. This implies that there is little benefit in selecting cultivars with greater amounts of nonstructural carbohydrate stored at anthesis.

This has been supported by a study with maize selections of different heights having different amounts of stem nonstructural

FIGURE 3.3. Dry Weights of Ears (Uppermost), Stems (Middle), and Leaf Laminae (Bottom) for Control Plants and Plants Shaded During Grain Filling of Maize Hybrid B73 × Mo17 in 1988 in Texas

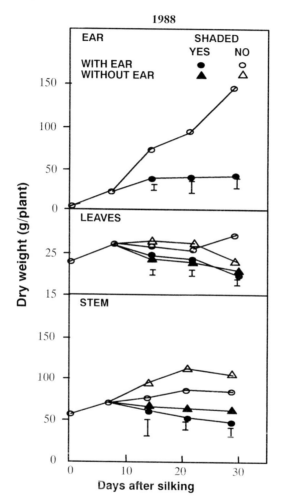

Source: Kiniry et al. (1992, p. 132).

Note: Bars are the mean standard errors (SE) for a harvest date. Shaded plants are the filled symbols. Plants with ears are the circles, and plants without ears are the triangles.

carbohydrate per plant (Edmeades and Lafitte, 1993). Shortened plants had 19 g less stem nonstructural carbohydrate than the tall plants at silking. However, using defoliation and planting density treatments, Edmeades and Lafitte found no evidence that this reduced nonstructural carbohydrate per plant caused a reduction in tolerance to stress during grain filling.

MECHANISMS OF SEED WEIGHT DETERMINATION IN MAIZE

Whereas sorghum generally increases individual seed dry weight when seed number is reduced (Kiniry, 1988), maize has frequently been found to show no significant change in seed dry weight due to changes in seed number (Tollenaar and Daynard, 1978; Jones and Simmons, 1983; Zink and Michael, 1985). This trend is particularly stable for those environments in which low temperatures (< 19°C) during grain filling do not limit biomass partitioning to the grains (Wilson, Muchow, and Murgatroyd, 1995). For maize grown at high latitudes, both air temperature and incident solar radiation are low during the grain-filling period (Otegui and Bonhomme, 1998; Maddonni, Otegui, and Bonhomme, 1998), affecting biomass production and its partitioning (see Figure 3.4) and, perhaps, sink activity (Ou-Lee and Setter, 1985). Under these conditions, grain yield can be limited by seed dry weight. In temperate environments, air temperature is usually above the thresholds that may affect radiation use efficiency (16°C) and biomass partitioning to grains (Otegui, Ruiz, and Petruzzi, 1996; Maddonni, Otegui, and Bonhomme, 1998), resulting in very stable seed dry weights (see Figure 3.4). Nevertheless, this stability of seed dry weight needed to be tested across genotypes (Maddonni, Otegui, and Bonhomme, 1998) and methods of reducing the number of seeds (Kiniry et al., 1990).

Genotypes found to have the ability to compensate for reductions in seed number may be more desirable in environments having greater probability of reduced seeds per plant due to water stress near silking. The objective of our field study (Kiniry et al., 1990) was to investigate the response of seed dry weight to reduced number of seeds. Different methods were used to reduce seeds per ear. A

FIGURE 3.4. Response of Individual Seed Dry Weight (SDWT) to Plant Weight Gain per Seed During the Effective Grain-Filling Period

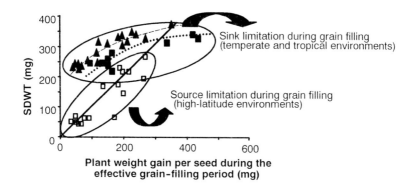

Source: Adapted from Maddonni, Otegui, and Bonhomme (1998, p. 262) and Cirilo and Andrade (1996, p. 329).

Note: Closed symbols are for the temperate region of Argentina, and open symbols for the high latitude in France. In Argentina, triangles identify data from Balcarce, and squares, from Salto-Rojas. The solid line represents the 1:1 relation. Dotted (Salto-Rojas) and dashed (Balcarce) lines are the fitted models.

number of cultivars of different seed types were included to test the generality of the results.

Seeds per plant were reduced by controlled pollination (bagging ears) at different dates after silking. We used maize hybrid B73 × Mo17, a white popcorn hybrid, and five open-pollinated cultivars with different seed types in an attempt to investigate the generality of any seed weight compensation. Ears were bagged at three, five, seven, or nine days after silking in 1987 and two or four days after silking in 1988. Seeds per ear were reduced by as much as 42 percent in the earliest treatment.

Hybrid B73 × Mo17 had a 19 to 25 percent increase in mean seed dry weight when seeds per plant were reduced by 26 to 45 percent. A popcorn hybrid showed no response of seed dry weight to these treatments. Similar differences in the response were found for diverse seed types when the ears were bagged at three to nine days after silking.

The increase in seed dry weight in response to decreased seed number was not due simply to elimination of smaller, more apical seeds. Weight of twenty seeds in the basal portion of the ear showed the same trends of weight increase in 1988 as did overall mean seed dry weight. Weights of twenty seeds for B73 × Mo17 were significantly greater than the control for the two- and four-day treatments, respectively. The percentage increases in weight of twenty seeds were similar to the corresponding percentage increase in overall seed dry weight for this hybrid. Likewise, there was a trend of increased weight of twenty seeds for the two-day treatment of Gourdseed, similar to the increase in overall seed dry weight for the treatment.

The relative increase in seed volume of these basal seeds corresponded closely to the increase in weight per twenty seeds. The two- and four-day treatments of B73 × Mo17 were significantly greater than the control, within 1 percent of the corresponding changes in weight per twenty seeds. Though both seed volume and weight per twenty seeds of bagging treatments were not significantly different from the control for Gourdseed, the trend for both treatments was a 3 percent increase in seed volume and a 4 to 5 percent increase in weight per twenty seeds, with respect to the control.

Thus, we found a response of seed dry weight to reduced seeds per ear and found differences among hybrids in their responsiveness. Evidence from this study and others implies that the developing pericarp may sometimes constrain final seed dry weight. Early controlled pollination treatments of B73 × Mo17 had increases in seed volume similar to the increase in seed dry weight. B73 × Mo17 has a thinner pericarp than most popcorn cultivars (Tracy and Galinat, 1987). The thicker pericarp of popcorn may physically restrain its seed size.

CONCLUSIONS

1. Final seed number per plant is linearly dependent on photosynthesis and light interception soon after silking. Responsiveness varies among genotypes. Plant types producing greater seeds per unit intercepted light can be chosen for improving yield,

especially under high management inputs and high planting densities.

2. Increased storage of carbohydrate at anthesis does not appear to be a likely candidate for increased grain yields, even in conditions of stress-limited carbohydrate supply late in grain filling. These stored assimilates were used for maintaining the vegetative organs of the plant, with minimal benefit to ear growth under severe shading stress.

3. Maize shows variability among genotypes in the ability to increase weight per seed when short-term stress near silking reduces final grain number. Thus, this may be a trait to be studied for use in selecting hybrids more capable of compensating for stress near silking.

REFERENCES

Andrade, F.H., Cirilo, A.G., and Echarte, L. (2000). Factors affecting kernel number in maize. In Otegui, M.E. and Slafer, G.A. (Eds.), Physiological bases for maize improvement (pp. 59-73). Binghamton, NY: Food Products Press.

Andrade, F.H., Cirilo, A.G., Uhart, S.A., and Otegui, M.E. (1996). *Ecofisiología del cultivo de maíz*. Balcarce, Argentina: La Barrosa, CERBAS, and Dekalb Press.

Andrade, F.H., Otegui, M.E., and Vega, C. (2000). Intercepted radiation at flowering and kernel number in maize. *Agronomy Journal* 92: 92-97.

Andrade, F.H., Uhart, S.A., and Frugone, M.I. (1993). Intercepted radiation at flowering and kernel number in maize: Shade versus plant density effects. *Crop Science* 33: 482-485.

Andrade, F.H., Vega, C., Uhart, S.A., Cirilo, A.G., Cantarero, M., and Valentinuz, O. (1999). Kernel number determination in maize. *Crop Science* 39: 453-459.

Bolaños, J. and Edmeades, G.O. (1993). Eight cycles of selection for drought tolerance in lowland tropical maize. II. Responses in reproductive behavior. *Field Crops Research* 31: 253-268.

Bonhomme, R., Derieux, M., Duburcq, J.B., and Ruget, F. (1984). Variation in ovule number at silking in various corn genotypes. *Maydica* 29: 101-107.

Boyle, M.G., Boyer, J.S., and Morgan, P.W. (1991). Stem infusion of liquid culture medium prevents reproductive failure of maize at low water potential. *Crop Science* 31: 1246-1252.

Campbell, D.K. and Hume, D.J. (1970). Evaluation of a rapid technique for measuring soluble solids in corn stalks. *Crop Science* 10: 625-626.

Cirilo, A.G. and Andrade, F.H. (1994). Sowing date and maize productivity. II. Kernel number determination. *Crop Science* 34: 1044-1046.

Cirilo, A.G. and Andrade, F.H. (1996). Sowing date and kernel weight in maize. *Crop Science* 36: 325-331.

Daynard, T.B., Tanner, J.W., and Hume, D.J. (1969). The contribution of stalk soluble carbohydrates to grain yield in corn (*Zea mays* L.). *Crop Science* 9: 831-834.

DuPlessis, D.P. and Dijkhuis, F.J. (1967). The influence of the time lag between pollen-shedding and silking on the yield of maize. *South African Journal of Agricultural Science* 10: 667-674.

Edmeades, G.O., Bolaños, J., Hernandez, M., and Bello, S. (1993). Cause for silk delay in a lowland tropical maize population. *Crop Science* 33: 1029-1035.

Edmeades, G.O. and Lafitte, H.R. (1993). Defoliation and plant density effects on maize selected for reduced plant height. *Agronomy Journal* 85: 850-857.

Fischer, R.A. (1985). Number of kernels in wheat crops and the influence of solar radiation and temperature. *Journal of Agricultural Science, Cambridge* 105: 447-461.

Grant, R.F., Jackson, B.S., Kiniry, J.R., and Arkin, G.F. (1989). Water deficit timing effects on yield components in maize. *Agronomy Journal* 81: 61-65.

Hall, A.J., Lemcoff, J.H., and Trápani, N. (1981). Water stress before and during flowering in maize and its effects on yield, its components, and their determinants. *Maydica* 26: 19-38.

Hall, A.J., Vilella, F., Trápani, N., and Chimenti, C. (1982). The effects of water stress and genotype on the dynamics of pollen-shedding and silking in maize. *Field Crops Research* 5: 349-363.

Jones, R.J. and Simmons, S.R. (1983). Effect of altered source-sink ratio on growth of maize kernels. *Crop Science* 23: 129-134.

Kiniry, J.R. (1988). Mechanism of kernel weight increase in response to decreased kernel number in grain sorghum. *Agronomy Journal* 80: 221-226.

Kiniry, J.R. and Knievel, D.P. (1995). Response of maize seed number to solar radiation intercepted soon after anthesis. *Agronomy Journal* 87: 228-234.

Kiniry, J.R. and Ritchie, J.T. (1985). Shade-sensitive interval of kernel number of maize. *Agronomy Journal* 77: 711-715.

Kiniry J.R., Tischler, C.R., Rosenthal, W.D., and Gerik, T.J. (1992). Nonstructural carbohydrate utilization by sorghum and maize shaded during grain growth. *Crop Science* 32: 131-137.

Kiniry, J.R., Wood, C.A., Spanel, D.A., and Bockholt, A.J. (1990). Seed weight response to decreased seed number in maize. *Agronomy Journal* 82: 98-102.

Lafitte, H.R. and Edmeades, G.O. (1995). Stress tolerance in tropical maize is linked to constitutive changes in ear growth characteristics. *Crop Science* 35: 820-826.

Maddonni, G., Otegui, M.E., and Bonhomme, R. (1998). Grain yield components in maize. II. Postsilking crop growth and kernel weight. *Field Crops Research* 56: 257-264.

McBee, G.G., Waskom III, R.M., Miler, F.R., and Creelman, R.A. (1983). Effect of senescence and nonsenescence on carbohydrates in sorghum during late kernel maturity states. *Crop Science* 23: 372-376.

McPherson, H.G. and Boyer, J.S. (1977). Regulation of grain yield by photosynthesis in maize subjected to a water deficiency. *Agronomy Journal* 69: 714-718.

Otegui, M.E. (1995). Prolificacy and grain yield components in modern Argentinean maize hybrids. *Maydica* 40: 371-376.

Otegui, M.E. (1997). Kernel set and flower synchrony within the ear of maize. II. Plant population effects. *Crop Science* 37: 448-455.

Otegui, M.E., Andrade, F.H., and Suero, E.E. (1995). Growth, water use, and kernel abortion of maize subjected to drought at silking. *Field Crops Research* 40: 87-94.

Otegui, M.E. and Bonhomme, R. (1998). Grain yield components in maize. I. Ear growth and kernel set. *Field Crops Research* 56: 247-256.

Otegui, M.E. and Melón, S. (1997). Kernel set and flower synchrony within the ear of maize. I. Sowing date effects. *Crop Science* 37: 441-447.

Otegui, M.E., Ruiz, R., and Petruzzi, D. (1996). Modeling hybrid and sowing date effects on potential grain yield of maize in a humid temperate region. *Field Crops Research* 47: 167-174.

Ou-Lee, T.-M. and Setter, T.L. (1985). Effect of increased temperature in apical regions of maize ears on starch-synthesis enzymes and accumulation of sugars and starch. *Plant Physiology* 79: 852-855.

Penning de Vries, F.W.T., van Laar, H.H., and Chardon, M.C.M. (1993). Bioenergetics of growth of seeds, fruits, and storage organs. In Yoshida, I. (Ed.), *Potential productivity of field crops under different environments* (pp. 37-59). Los Baños, Philippines: International Rice Research Institute.

Ruget, F. (1993). Contribution of storage reserves during grain filling of maize in northern European conditions. *Maydica* 38: 51-59.

Schussler, J.R. and Westgate, M.E. (1994). Increasing assimilate reserves does not prevent kernel abortion at low water potential in maize. *Crop Science* 34: 1569-1576.

Schussler, J.R. and Westgate, M.E. (1995). Assimilate flux determines kernel set at low water potential in maize. *Crop Science* 35: 1074-1080.

Swank, J.C., Below, F.E., Lambert, R.J., and Hageman, R.H. (1982). Interaction of carbon and nitrogen metabolism in productivity of maize. *Plant Physiology* 70: 1185-1190.

Tollenaar, M. and Daynard, T.B. (1978). Dry weight, soluble sugar content, and starch content of maize kernels during the early post-silking period. *Canadian Journal of Plant Science* 58: 199-206.

Tollenaar, M., Dwyer, L.M., and Stewart, D.W. (1992). Ear and kernel formation in maize hybrids representing three decades of grain yield improvement in Ontario. *Crop Science* 32: 432-438.

Tracy, W.F. and Galinat, W.C. (1987). Thickness and cell layer number of the pericarp of sweet corn and some of its relatives. *HortScience* 22: 645-647.

Vanderlip, R.L., Charles-Edwards, D.A., Foale, M.A., and Ferraris, R. (1984). Predicting tiller number and seed number in grain sorghum. *Agronomy Abstracts*, p. 138.

Westgate, M.E. and Boyer, J.S. (1986). Reproduction at low silk and pollen water potentials in maize. *Crop Science* 26: 951-956.

Williams, W.A., Loomis, R.S., Duncan, W.G., Dovrat, A., and Nuñez, A.F. (1968). Canopy architecture at various population densities and the growth and grain yield of corn. *Crop Science* 8: 303-308.

Wilson, D.R., Muchow, R.C., and Murgatroyd, C.J. (1995). Model analysis of temperature and solar radiation limitations to maize potential productivity in a cool climate. *Field Crops Research* 43: 1-18.

Zink, F. and Michael, G. (1985). The effect of reduced number of kernels and of leaves per plant on the nitrogen storage characteristics of the kernels of two maize hybrids with different grain N content. *Zeitschrift für Acker- und Pflanzenbau (Journal of Agronomy and Crop Science)* 154: 203-212.

Zinselmeier, C., Lauer, M.J., and Boyer, J.S. (1995). Reversing drought-induced losses in grain yield: Sucrose maintains embryo growth in maize. *Crop Science* 35: 1390-1400.

Chapter 4

Improving Maize Grain Yield Potential in the Tropics

Russell C. Muchow

INTRODUCTION

Potential productivity, defined as the yield obtained under high-input conditions when all manageable factors are controlled, varies across agroecologies, principally in response to the solar radiation and temperature regime. Such high-input agriculture is important to world food supplies, but, increasingly, such production systems need to match better resource supply with resource demand so that resources (e.g., water and nutrients) are not lost from the production system, with potential negative impacts on the environment. Having knowledge of potential productivity can assist in identifying opportunities to improve yield and resource utilization in diverse production systems.

To advance this analysis, it is necessary to understand the hierarchy of production levels, as shown in Table 4.1. Climate is a key determinant of potential yield, with attainable yield also being determined by the availability of water and nutrients. The actual yield measured in farmers' fields is also subject to the confounding influences of pests, diseases, and weeds; timing of management operations; waterlogging; soil impediments (e.g., compaction, salinity, acidity, etc.); and calamities (e.g., cyclones and floods). In addition, it may not be economical to operate at maximum production levels, or environmental considerations may limit the management of water and nutrients. Social factors, such as the availability of labor at key times and lifestyle decisions, also have an impact on actual yields. In

TABLE 4.1. Hierarchy of Production Levels

Production Level	Determinants
Potential yield	Temperature, radiation
Attainable yield	Potential yield, water and nutrient supply
Actual yield	Attainable yield, crop management, pests and diseases
	Soil impediments, lodging, calamities
	Economic, environmental, and social factors

tropical environments, in particular, these limitations to potential yield are often more important than sole consideration of the biophysical environment. Hence, it is important in analyzing potential yields to be aware of the relationship between potential yields and actual yields observed in commercial production.

In addressing the subject of improving maize grain yield potential in the tropics, this chapter analyzes (1) the determinants of potential yield in terms of radiation interception and use; (2) variation in benchmark potential maize yields in tropical, subtropical, and temperate environments and the contributing factors; and (3) opportunities for further increases in maize yield potential.

DETERMINANTS OF POTENTIAL YIELD

For a given variety, the two factors that determine crop yield are environment and management. Quantitatively, these impacts on crop yield can be described in terms of the interception and use of solar radiation. For many crop species grown under well-watered conditions and ample nutrition in the absence of pests and disease, biological yield (biomass production) has been shown to be linearly related to the amount of radiation intercepted by the crop canopy (Williams, Loomis, and Lepley, 1965; Monteith, 1977). This relationship sets a finite limit on yield potential (Sinclair, 1994). The slope of the linear relationship, often termed the radiation use efficiency (RUE), is curvilinearly related to light-saturated leaf carbon dioxide (CO_2) assimilation rate, C_A, and at current maximum RUE values, very large increases in C_A would be required to achieve even modest further increases in maximum RUE (Sinclair and Horie, 1989). Gifford (1974) reported that canopy biomass accumulation is relatively insensitive to increases in C_A, and in a review, Gifford and colleagues (1984) concluded that efforts to increase crop growth rates through increasing the rate of net photosynthesis have met with little success for a number of crop species. Similarly, Dwyer and Tollenaar (1989) found no significant differences between new and old maize hybrids in potential leaf photosynthesis. Of more importance in realizing yield potential is the maintenance of high rates of C_A throughout the total growing season. This is particularly so during grain filling, where a decline in C_A is often related to inadequate nitrogen (N) supply (Kropff et al., 1993; Muchow and Sinclair, 1994). Likewise, the prevention of abiotic and biotic stresses is important in raising yields toward the potential, and the narrowing of this gap between actual and potential yields should be the focus of crop improvement programs.

That portion of the biological yield which is harvested is referred to as the economic yield. The proportion of the biological yield that is present as economic yield is referred to as the harvest index (HI). Evans (1993) reviewed the major changes in crops that have been brought about by coselection for high yield and suitability for modern, often high-input husbandry and concluded that the increase in HI has been one of the most significant advances in crop yield potential. Over the past thirty years, for example, the HI of small-grain cereals has increased from 0.3 to 0.5. In many crop species, HI under potential growth conditions is above 0.5, and an important question is how much more can harvest index be increased. In intensively bred cereals grown under high-input conditions, Austin and colleagues (1980) estimated a maximum possible HI of 0.63, but this is unlikely, given that shorter and weaker stems increase the susceptibility to lodging. While HI is relatively conservative under high-input conditions, it can be substantially reduced under severe water and nitrogen stress (Muchow 1988, 1989b). Sinclair (1988) reported that biomass is more sensitive to stress than is HI, with HI decreasing only when biomass production is less than 40 percent of potential biomass production. There is scope under stress conditions to increase HI toward the potential, and several studies have shown such improvements (Fischer, Edmeades, and Johnson, 1989; Tollenaar, McCullough, and Dwyer, 1993).

Given that there appears to be relatively little scope in increasing the maximum efficiency of radiation use in biomass production, nor in increasing the partitioning of that biomass to economic yield under potential growth conditions, high yield will be obtained where radiation interception is maximized and RUE is maintained at maximum values.

Radiation interception is strongly dependent on growth duration, as determined by crop phenology, and on leaf canopy development. Leaf canopy development, as influenced by ambient temperature, determines the leaf area index of the crop, thereby dictating the proportion of the incident radiation that is intercepted (Muchow and Carberry, 1989). The duration of crop growth sets the maximum time that the incident radiation (which varies with location, year, and season) can be intercepted. For many cropping systems, cool and freezing temperatures effectively define the growth duration. Similarly, in monsoonal rain-fed cropping systems, lack of water at the onset of the dry season effectively defines the length of the growing season. In maize, where the reproductive growth comprises the economic yield, the durations of vegetative and reproductive growth, as modified by temperature and photoperiod (Muchow and Carberry, 1989), also set a limit to the length of the growing season. Radiation interception can also be reduced by stress factors, with water and N supply having a major influence (Novoa and Loomis, 1981; Muchow, 1989b; Muchow, 1994).

Recently, Sinclair and Muchow (1998) did a comprehensive review of RUE in different crop species. After taking into account the different methods of measurement and calculation of RUE, they concluded that the maximum RUE for maize is consistently in the range of 1.6 to 1.7 grams per megajoule $(g \cdot MJ^{-1})$ during vegetative growth, and that the RUE during reproductive growth decreases such that the seasonal RUE ranges from 1.3 to 1.7 $g \cdot MJ^{-1}$. This was based on the experimental observations of Muchow and Davis (1988); Muchow (1989a); Daughtry and co-workers (1992); Tollenaar and Aguilera (1992); Muchow (1994); Kiniry (1994); and Otegui and colleagues (1995).

Radiation use efficiency under potential growth conditions is depressed at supraoptimal and suboptimal temperatures. For example, in maize, maximum RUE occurs at mean daily temperatures from 20 to 40°C and decreases to zero at a base temperature of 8°C and an extreme mean temperature of 50°C (Carberry, Muchow, and McCown, 1989). Andrade and co-workers (1992) concluded that low RUE values for maize grown in Balcarce, Argentina, were the result of low temperature. In additional experiments with maize, using varying sowing dates, Andrade, Uhart, and Cirilo (1993) found a linear decrease in RUE associated with a decrease in mean temperature from 21 to 16°C. They found variation in RUE during vegetative growth of 1.05 to 1.52 $g \cdot MJ^{-1}$ over five years that was associated with yearly differences in temperature.

Vapor pressure deficit has been indicated as having the potential for influencing RUE (Stockle and Kiniry, 1990; Kiniry et al., 1998). To the extent that a large vapor pressure deficit may result in decreased photosynthetic rates, then a decrease in RUE would be expected. The influence of vapor pressure deficit on leaf photosynthetic rate tends, however, to develop at fairly high vapor pressure deficits (greater than 2 kilopascals [kPa]), and, even then, the decreases in photosynthetic rate are fairly modest (Stockle and Kiniry, 1990). Nevertheless, Stockle and Kiniry (1990) presented regressions for maize using a number of data sources for RUE showing a decrease in RUE with increasing vapor deficit. Their analysis indicated a large effect that was greater than predicted by leaf behavior. Since decreasing vapor pressure deficit is likely to be associated with a number of other environmental variables (e.g., level of solar radiation and fraction of diffuse radiation), a simple regression of RUE against vapor pressure deficit is confounded (Sinclair and Muchow, 1998). In contrast to this, Muchow and Sinclair (1994) found no variation in RUE with vapor pressure deficit for maize across six environments of differing vapor pressure deficit. Considering the reasonable stability observed in most crops across environments, it seems likely that vapor pressure deficit has only a small influence on RUE in most cropping situations.

Radiation use efficiency is decreased by water and nitrogen stress (Muchow, 1989b, 1994). However, even under potential growth conditions, it is difficult to maintain leaf N values sufficiently high for the whole growth duration due to mobilization of leaf N to grain N, with a subsequent decrease in RUE during grain filling (Kropff et al., 1993; Muchow and Sinclair, 1994). Even though maize produces grain with relatively low grain N concentration (Muchow, 1998), the N required to support grain growth is substantial. This N comes from both direct soil N uptake and N mobilization from leaves and stems. Even under high fertilizer rates and ample water supply (i.e., potential growth conditions), the grain N demand can be sufficiently high for leaf N levels to fall, resulting in RUE below maximum values. This is evidenced in the study by Muchow (1994) in which maize was grown under high water and N supply. Here, the maximum RUE for seven sowings in tropical and subtropical environments was 1.61 $g \cdot MJ^{-1}$, yet the average RUE from sowing to maturity was 1.34 $g \cdot MJ^{-1}$. Hence, N can limit the physiological performance of maize even under potential growth conditions.

BENCHMARKING MAIZE YIELD POTENTIAL

Knowledge of the benchmark yield for a given cultivar, location, and season allows assessment of how far current yields are below the potential yield and what is the scope for yield improvement. Muchow, Sinclair, and Bennett (1990) examined the effects of variation in solar radiation and temperature on the potential yield of maize, based on field experiments conducted at different locations, ranging from latitude 14°S to 40°N. Maize hybrids of similar phenology (eighteen leaves) were selected, and a subset of this analysis is shown in Table 4.2. Grain yields were more than double (19 tonnes per hectare [$t \cdot ha^{-1}$]) in the high-altitude temperate (Grand Junction) environment compared to the tropical (Katherine) environment (9.4 $t \cdot ha^{-1}$). In the tropics, the growth duration was much shorter than in higher latitudes, but at a given location, potential grain yield was not necessarily related to growth duration. For example, there was little difference in grain yield for the two Katherine sowings shown in Table 4.2, as the higher incident radiation compensated for the shorter growth duration.

To understand the variation in yield across locations, Muchow, Sinclair, and Bennett (1990) developed a simple mechanistic growth model. Crop phenology and leaf growth were calculated from daily mean temperature, based on functions derived from the experimental data of Muchow and Carberry (1989). Daily biomass accumulation was calculated by estimating the amount of radiation intercepted and assuming a maximum radiation use efficiency of 1.6 g · MJ^{-1}, based on the experimental observations of Muchow and Davis (1988). Grain growth was simulated using a linear increase in harvest index to a maximum value of 0.5 during grain filling, based on the experimental observations of Muchow (1990). Field experimentation at Katherine, in tropical Australia, provided all the quantitative functions in the model, and when the model was run with inputs of daily maximum and minimum temperatures and solar radiation from the experiments at the different locations (in Australia and the United States), there was close agreement between observed and simulated grain yield data (see Table 4.2).

Table 4.2. Effect of Temperature and Radiation Regime on Observed (Obs.) and Simulated (Sim.) Grain Yield (at 15.5 Percent Moisture) and Growth Duration from Emergence to Maturity for Maize Cultivars at Different Locations

Location (latitude)	Sowing date and cultivar	Duration days	Obs. grain yield t·ha^{-1}	Sim. grain yield t·ha^{-1}	Mean daily tempera-ture °C	Incident radiation MJ·m^{-2} per day
Katherine, NT, AUS. 14° 28′S	30 Aug. 86 Dekalb XL82	84	9.47	9.60	28.7	25.5
Katherine, NT, AUS. 14° 28′S	6 Feb. 85 Dekalb XL82	95	9.34	9.36	26.3	22.2
Lawes, Q AUS. 27° 34′S	3 Oct. 90 Dekalb XL82	118	12.2	11.2	24.8	22.1
Champaign, IL USA 40° 7′N	4 May 82 Agway 849X	126	12.7	12.9	21.5	19.9
Grand Junction, CO 39° 4′N	7 May 84 Funk G4507	138	19.0	19.3	19.8	28.3
Grand Junction, CO 39° 4′N	28 Apr. 86 Dekalb 656	153	17.1	17.1	18.0	22.4

Source: Data from Muchow, Sinclair, and Bennett (1990), except for Lawes, in Muchow (1994).

The model analysis showed that both temperature and solar radiation influence the variation in potential maize yield across environments. The primary influence of temperature was on growth duration, with lower temperatures increasing the time that the crop can intercept radiation. Biomass accumulation is directly proportional to the amount of radiation intercepted, and for a given harvest index, grain yield is directly proportional to biomass. Consequently, high maize yield was associated with low temperature and high solar radiation within the range of environments studied (see Table 4.2). Furthermore, this analysis indicates that the solar radiation and temperature regime set a finite limit to potential yield in a given environment.

OPPORTUNITIES FOR IMPROVEMENT IN MAIZE GRAIN YIELD POTENTIAL

While the analysis thus far has shown that temperature and radiation regime set a finite limit to yield potential, the attainment of potential productivity requires high resource inputs. The alleviation of water deficit is easily understood as important to the attainment of potential productivity. However, crop N status is frequently a limitation to the attainment of potential productivity, even under apparent high N supply. As discussed earlier, high leaf N is required to maximize RUE and productivity, especially during grain filling (Muchow and Sinclair, 1994).

Sinclair and Muchow (1995) developed a maize growth simulation model whereby physiological activity was calculated based on N levels in the crop tissue. A sensitivity test performed with the model to investigate the importance of various physiological parameters on the determination of grain yield showed that under N-limiting conditions, grain yield was most sensitive to the minimum grain N concentration. This is a consequence of grain N demand not being able to be met by soil N uptake, even under high N supply, with mobilization of leaf N to the grain and consequent loss of photosynthetic capacity, as expressed in RUE during grain filling (Muchow, 1994; Muchow and Sinclair, 1994). An approach for improving grain yield and the efficiency of N utilization might be to genetically decrease the minimum grain N concentration. However, lowered protein content of the grain may make it less desirable as a food or feed.

It is clear from this analysis that it is essential to match the seasonal pattern of N supply to the N demand of the crop at each stage of development to achieve full yield potential. However, N is mobile in the soil system, and with the required high rates of N application, the possibility exists for losses (e.g., leaching), and negative environmental impact. Environmental considerations to minimize nutrient losses from the production systems may, in fact, limit yield potential because of the challenge of matching high inputs with crop demand, both temporally and spatially.

Simulation modeling has a role to play in defining both the crop demand and the potential losses from the system under different climatic conditions. Keating, Godwin, and Watiki (1991) have reviewed the range of models available to describe the effect the interaction of N supply with climatic factors has on yield. However, further development of models is required to simulate the crop-soil system, to assist in the development of strategies that best match resource supply with resource demand, with minimal loss of resources to the environment. This remains, perhaps, the biggest challenge to sustaining potential productivity in the cropping system.

CONCLUSIONS

Tropical environments pose a particular challenge to the attainment of potential productivity. First, high temperature accelerates crop development, except at higher altitudes within these environments, and reduces the potential amount of incident radiation that can be intercepted by the crop. Second, high temperature also increases pest and disease pressure in tropical environments, making it difficult to attain potential growth conditions. Third, potential productivity is dependent on high resource inputs, with water and nutrients frequently limiting productivity in tropical environments. Even where there are ample water and nutrient supplies, the ability of the maize plant to maintain adequate leaf area for full radiation interception during grain filling, and to maintain leaf N status so that RUE is maximized, is a constraint to potential productivity. While opportunities to increase inherent crop maximum photosynthetic capacity are limited, there are opportunities to improve N utilization during the life cycle. This is likely to lead to increases in potential biomass production (through maintenance of RUE during grain filling) and higher harvest index (through satisfying grain N demand).

Perhaps the greatest opportunity to increase maize grain yield potential in the tropics is to reduce the gap between actual and potential yields. Many nonphysiological factors result in poor yield potential (see Table 4.1). The challenge for physiologists and agronomists is to understand the yield potential in a given environment, and its variability across seasons and locations, and to match resource inputs to achieve such yield potential. Quantitative understanding of the determinants of yield and the enhancement of modeling tools that utilize that understanding are essential in improving maize grain yield potential in diverse production systems.

REFERENCES

Andrade, F.H., Uhart, S.A., Arguissain, G.G., and Ruiz, R.A. (1992). Radiation use efficiency of maize grown in a cool area. *Field Crops Research* 28: 345-354.

Andrade, F.H., Uhart, S.A., and Cirilo, A.G. (1993). Temperature affects radiation use efficiency in maize. *Field Crops Research* 32: 17-25.

Austin, R.B., Bingham, J., Blackwell, R.D., Evans, L.T., Ford, M.A., Morgan, C.L., and Taylor, M. (1980). Genetic improvements in winter wheat yields since 1990 and associated physiological changes. *Journal of Agricultural Science, Cambridge* 94: 675-689.

Carberry, P.S., Muchow, R.C., and McCown, R.L. (1989). Testing the CERES-maize simulation model in a semi-arid tropical environment. *Field Crops Research* 20: 297-315.

Daughtry, C.S.T., Gallo, K.P., Goward, S.N., Prince, S.D., and Kustas, W.P. (1992). Spectral estimates of absorbed radiation and phytomass production in corn and soybean canopies. *Remote Sensing of Environment* 39:141-152.

Dwyer, L.M. and Tollenaar, M. (1989). Genetic improvement in photosynthetic response of hybrid maize cultivars, 1959 to 1988. *Canadian Journal of Plant Science* 69: 81-91.

Evans, L.T. (1993). Processes, genes, and yield potential. In Buxton, D.R., Shibles, R., Forsberg, R.A., Blad, B.L., Asay, K.H., Paulsen, G.M., and Wilson, R.F. (Eds.), *International crop science*, Volume I (pp. 687-701). Madison, WI: CSSA.

Fischer, K.S., Edmeades, G.O., and Johnson, E.C. (1989). Selection for the improvement of maize yield under moisture deficits. *Field Crops Research* 22: 227-243.

Gifford, R.M. (1974). A comparison of potential photosynthesis, productivity and yield of plant species with differing photosynthetic metabolism. *Australian Journal of Plant Physiology* 1: 107-117.

Gifford, R.M., Thorne, J.H., Hitz, W.D., and Gianquinta, R.T. (1984). Crop productivity and photoassimilate partitioning. *Science* 225: 801-808.

Keating, B.A., Godwin, D.C., and Watiki, J.M. (1991). Optimising nitrogen inputs in response to climatic risk. In Muchow, R.C. and Bellamy, J.A. (Eds.), *Climatic risk in crop production: Models and management for the semiarid tropics and subtropics* (pp. 329-358). Wallingford, UK: CAB International.

Kiniry, J.R. (1994). Radiation-use efficiency and grain yield of maize competing with johnsongrass. *Agronomy Journal* 86: 554-557.

Kiniry, J.R., Landivar, J.A., Witt, M., Gerik, T.J., Cavero, J., and Wade, L.J. (1998). Radiation-use efficiency response to vapor pressure deficit for maize and sorghum. *Field Crops Research* 56: 265-270.

Kropff, M.J., Cassman, K.G., van Laar, H.H., and Peng, S. (1993). Nitrogen and yield potential of irrigated rice. *Plant and Soil* 155/156: 391-394.

Monteith, J.L. (1977). Climate and the efficiency of crop production in Britain. *Philosophical Transactions of the Royal Society of London, Britain* 281: 277-294.

Muchow, R.C. (1988). Effect of nitrogen supply on the comparative productivity of maize and sorghum in a semiarid tropical environment. III. Grain yield and nitrogen accumulation. *Field Crops Research* 18: 31-44.

Muchow, R.C. (1989a). Comparative productivity of maize, sorghum and pearl millet in a semi-arid tropical environment. I. Yield potential. *Field Crops Research* 20: 191-205.

Muchow, R.C. (1989b). Comparative productivity of maize, sorghum and pearl millet in a semi-arid tropical environment. II. Effect of water deficits. *Field Crops Research* 20: 207-219.

Muchow, R.C. (1990). Effect of high temperature on grain growth in field-grown maize. *Field Crops Research* 23: 145-158.

Muchow, R.C. (1994). Effect of nitrogen on yield determination in irrigated maize in tropical and subtropical environments. *Field Crops Research* 38: 1-13.

Muchow R.C. (1998). Nitrogen utilization efficiency in maize and grain sorghum. *Field Crops Research* 56: 209-216.

Muchow, R.C. and Carberry, P.S. (1989). Environmental control of phenology and leaf growth in a tropically adapted maize. *Field Crops Research* 20: 221-236.

Muchow, R.C. and Davis, R. (1988). Effect of nitrogen supply on the comparative productivity of maize and sorghum in a semi-arid tropical environment. II. Radiation interception and biomass accumulation. *Field Crops Research* 18: 7-30.

Muchow, R.C. and Sinclair, T.R. (1994). Nitrogen response of leaf photosynthesis and canopy radiation use efficiency in field-grown maize and sorghum. *Crop Science* 34: 721-727.

Muchow, R.C., Sinclair, T.R., and Bennett, J.M. (1990). Temperature and solar radiation effects on potential maize yield across locations. *Agronomy Journal* 82: 338-343.

Novoa, R. and Loomis, R.S. (1981). Nitrogen and plant production. *Plant Soil* 58: 177-204.

Otegui, M.E., Nicolini, M.G., Ruiz, R.A., and Dodds, P.A. (1995). Sowing date effects on grain yield components for different maize genotypes. *Agronomy Journal* 87: 29-33.

Sinclair, T.R. (1988). Selecting crops and cropping systems for water-limited environments. In Bidinger, F.R. and Johansen, C. (Eds.), *Drought research priorities for the dryland tropics* (pp. 87-94). Patancheru, India: ICRISAT.

Sinclair, T.R. (1994). Limits to crop yield? In Boote, K.J., Bennett, J.M., Sinclair, T.R., and Paulsen, G.M. (Eds.), *Physiology and determination of crop yield* (pp. 509-532). Madison, WI: ASA, CSSA, and SSSA.

Sinclair, T.R. and Horie, T. (1989). Leaf nitrogen, photosynthesis, and crop radiation use efficiency: A review. *Crop Science* 29: 90-98.

Sinclair, T.R. and Muchow, R.C. (1995). Effect of nitrogen supply on maize yield. I. Modeling physiological responses. *Agronomy Journal* 87: 632-641.

Sinclair, T.R. and Muchow, R.C. (1998). Radiation use efficiency. *Advances in Agronomy* 65: 215-265.

Stockle, C.O. and Kiniry, J.R. (1990). Variability in crop radiation-use efficiency associated with vapor-pressure deficit. *Field Crops Research* 25: 171-181.

Tollenaar, M. and Aguilera, A. (1992). Radiation use efficiency of an old and a new maize hybrid. *Agronomy Journal* 84: 536-541.

Tollenaar, M., McCullough, D.E., and Dwyer, L.M. (1993). Physiological basis of the genetic improvement of corn. In Slafer, G.A. (Ed.), *Genetic improvements in field crops: Current status and development* (pp. 183-236). New York: Marcel Dekker.

Williams, W.A., Loomis, R.S., and Lepley, C.R. (1965). Vegetative growth of corn as affected by population density. I. Productivity in relation to interception of solar radiation. *Crop Science* 5: 211-219.

Chapter 5

Factors Affecting Kernel Number in Maize

Fernando H. Andrade
Alfredo G. Cirilo
Laura Echarte

INTRODUCTION

In most cases, maize grain yield is more closely associated with kernel number than with kernel weight. For a specific genotype, the number of fertilized spikelets that survive and form mature kernels is more critical and more affected by environmental conditions than the total number of differentiated spikelets. Moreover, spikelet survival is correlated with the physiological condition of the crop at a period bracketing flowering.

Understanding the mechanisms of kernel number determination is of great importance to maize physiologists, modelers, and breeders. The number of kernels per plant (KNP) depends on the number of ears per plant and the number of kernels per ear that achieve physiological maturity. To optimize (maximize) this yield component, we must understand the mechanisms that control it at the critical period for grain number determination (CP). Those are plant growth rate (PGR), dry-matter partitioning to the ears, duration of the period, and number of kernels fixed per unit of biomass allocated to the ear (grain efficiency factor, GEF).

The first part of this chapter discusses the critical period for kernel number determination. The second part describes the relationship between KNP and PGR at a period bracketing flowering. This last variable is taken as an indicator of the physiological condition of the plant at this critical moment. The next section analyzes the KNP/PGR relationship for conditions in which nitrogen and water availability are variable. The fourth part considers also the duration of the critical period for kernel set and introduces the concepts of total crop growth during a specific period and growth per unit thermal time. Another section discusses other factors that affect kernel number, such as dry-matter partitioning to reproductive structures and the grain efficiency factor. Finally, a general model of kernel number determination in maize is presented. More emphasis is placed on those topics not extensively covered in a recent review by Otegui and Andrade (2000).

CRITICAL PERIOD
FOR KERNEL NUMBER DETERMINATION

Kernel number in maize is greatly affected by stress in the period between two weeks before to three weeks after silking (Claassen and Shaw, 1970; Tollenaar and Daynard, 1978; Fischer and Palmer, 1984; Andrade et al., 1996). Reductions in growth rate during this period have more impact on grain yield than similar reductions in other growth stages. Then, final kernel number per unit area is a function of radiation interception or crop growth rate during a period encompassing flowering (Tollenaar, Dwyer, and Stewart, 1992; Fischer and Palmer, 1984; Kiniry and Ritchie, 1985; Cirilo and Andrade, 1994). In general, restrictions during this critical period affect kernel set through reductions in both the number of ears per plant and the number of kernels per ear, with little effect on kernel weight (Fischer and Palmer, 1984; Kiniry and Ritchie, 1985; Aluko and Fischer, 1988).

The number of fertile ears per plant is set at flowering. Reduction of ear growth in sterile plants in relation to that of normal plants occurs just before or at silking; moreover, growth cessation of the second ear occurs during or just after silk emergence in the uppermost ear (Tollenaar, 1977).

The number of kernels per ear is determined in the near post-flowering period. This yield component is highly correlated with plant growth rate during the two- to three-week period after flowering (Cirilo and Andrade, 1994). Reductions in the number of viable kernels can continue up to two to three weeks after silking (Tollenaar, 1977; Fischer and Palmer, 1984; Cirilo and Andrade, 1994).

The decrease in the number of kernels per ear in response to shading at flowering is mainly due to kernel (or spikelet) abortion after silk emergence. Contrarily, the morphogenetic processes that determine rows per ear and kernels per row are relatively insensitive to variations in environmental and management conditions (Cirilo and Andrade, 1994; Uhart and Andrade, 1995). Mild water deficits (Otegui, Andrade, and Suero, 1995), sowing date (Otegui and Melón, 1997), and plant population (Otegui, 1997) did not affect the degree of floral differentiation reached at silking. Thus, for a specific genotype, kernel number per ear is a function of spikelet and kernel survival, and not of total number of differentiated spikelets. At a density of 8.5 plants per square meter (m^2), hybrid DK636 reached a maximum of 600 spikelets per ear a few days before flowering and a final kernel number of no more than 400 kernels per ear because of kernel and/or spikelet abortion. A recent review covers these aspects more extensively (Otegui and Andrade, 2000).

Then, grain yield in maize is closely correlated with kernel number, and this yield component is determined at a critical period bracketing silking. In the next sections, we will discuss the factors that determine kernel number during the critical period bracketing flowering, the duration of the period, dry-matter partitioning to the ears, and the number of kernels fixed per unit of total biomass allocated to the ear.

THE RELATIONSHIP BETWEEN KERNEL NUMBER AND PLANT GROWTH RATE

Plant growth rate at flowering is indicative of the physiological condition of the plant at the period bracketing silking and, thus, of the capacity of the plant to set kernels. Plant growth rate at CP is mainly affected by plant density, radiation level, temperature, water and nutrient availability, and genotype (Tollenaar, Dwyer, and Stewart, 1992; Andrade, Uhart, and Cirilo, 1993; Uhart and Andrade, 1995; Otegui, Andrade, and Suero, 1995; Andrade et al., 1999).

Figure 5.1 presents the relationship between KNP and PGR at the period bracketing silking for a specific hybrid. For these data, PGR changed because of plant density, radiation level, and year effects. The curvilinear relationship is characterized by (1) a PGR threshold value below which no kernels are set, (2) a PGR value above which increases in kernel number in response to increases in PGR are small, and (3) a PGR threshold value for prolificacy.

Kernel number per plant greatly responded to increases in PGR until about 4 grams per day (g/day). At PGRs greater than 4 g/day, increases in growth rates produced small increments in grains per uppermost ear because kernel number is approaching a maximum value. However, at rates greater than 6 g/day, a second grain-bearing ear could be set on the plant, providing plasticity in reproductive output (Prior and Russell, 1975).

Extrapolating the regression of the data of Figure 5.1, a positive x-intercept was found, indicating a threshold value close to 1 g/day for kernel set. Tollenaar, Dwyer, and Stewart (1992) also reported threshold values close to 1 g/day for hybrids grown in Canada.

A similar relationship was obtained with data from individual plants in which plant biomass was estimated through morphometric measurements (Vega 1997; Andrade et al., 1999). This threshold value for kernel set observed in maize is much higher than those found for other cultivated species (Vega, 1997). The high threshold value of PGR for kernel set is explained by the fact that the ear is a relatively weak sink around flowering. When PGR is low because of high density or poor environmental conditions, the ear suffers more than other parts of the plant. The delay in silk appearance relative to pollen shed (Tollenaar, 1977) and the benefits of male sterility for kernel set (Uhart et al., 1995) under stress conditions support this conclusion. More details about this relationship are presented in Andrade and colleagues (1999).

In conclusion, the curvilinear relationship between KNP and PGR at flowering can explain kernel number per plant when plant growth rate varies through plant density, incident radiation, or year effects. This relationship was obtained for maize grown without water and nutrient limitations. In the next section, the validity of this relationship for different levels of water and nitrogen availability is analyzed.

NITROGEN AND WATER

Unfavorable environmental conditions at flowering cause reductions in number of kernels per plant (Tollenaar, 1977; Fischer and Palmer, 1984; Kiniry and Ritchie, 1985; Jacobs and Pearson, 1991).

FIGURE 5.1. Relationship Between Kernel Number per Plant (or per Upper-most Ear) and Plant Growth Rate (PGR) at a Period Bracketing Silking

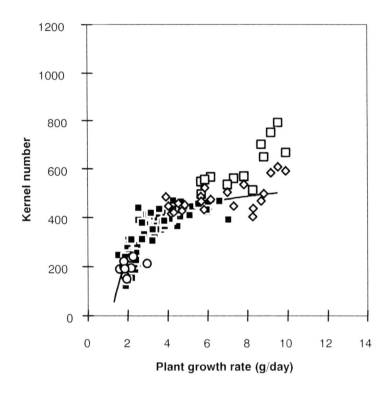

Source: Andrade et al. (1999, p. 456).

Note: Hybrid DK636 was sown in October and grown without nutrient and water limitations. PGR changed because of plant density, incident radiation, and year effects. Each point is the average of individual plants within each experimental unit. (◊), (o), and (■) indicate densities below 8, above 12, and between 8 and 12 plants/m², respectively. They indicate kernels per plant (KNP) when prolificacy was not observed in the experimental unit and kernels per uppermost ear when prolificacy was observed. When prolificacy was observed in the experimental unit, the sum of the kernels of the uppermost and second ear (average value of the plot) is indicated by (□). Standard errors were 17.5 kernels and 0.35 g/day for kernel number per plant and plant growth rate, respectively (calculated from pooled errors from different experiments). Data from all replications are shown. Inverse equations for uppermost ear only: KNP = 573−676/PGR (R^2 = 0.78).

In the previous section, a curvilinear relationship between PGR and KNP was reported for maize grown without water or nutrient deficiencies. The relationship between KNP and PGR must be validated for conditions in which plant growth at a period bracketing flowering varies because of changes in nutrient or water availability.

Nitrogen deficiencies decrease assimilate production at flowering and kernel set by reducing radiation interception and radiation use efficiency (Gifford et al., 1984; Uhart and Andrade, 1995). Moreover, much of the seed loss at low water potential could be accounted for by the lack of assimilate supply (Boyle, Boyer, and Morgan, 1991; Schussler and Westgate, 1991a,b). Thus, much of the nutritional and water stresses on kernel set would be explained through carbon assimilation and plant growth rate at flowering.

Czyzewicz and Below (1992) found that kernels ceased their growth when they were transferred from the field to in vitro culture without nitrogen and concluded that there is an obligatory nitrogen requirement to maintain growth and development of maize kernels after pollination. Similar results were reported by Below, Cazetta, and Seebauer (1998), who used stem infusion and in vitro culture to control and abruptly change the supply of carbon and nitrogen assimilates available for ear and kernel development. In relation to water stress, Schussler and Westgate (1991a,b) found that low ovary water potential could affect dry-matter partitioning to the ear by reducing ovary sink strength. Collectively, the data suggest that water and nitrogen have a direct role in reproductive development by controlling the ability of the kernel to utilize carbon. It is reasonable to expect direct effects of this essential nutrient on kernel set when its availability is low. These deficiencies could reduce dry-matter partitioning to the ear or the number of grains set per unit of dry matter allocated to the ear during the critical period for grain number determination.

If these more direct effects of water and nutrient availability on KNP were important, the curvilinear relationship between KNP and PGR obtained by varying PGR through incident radiation per plant would not be valid for conditions in which water and nutrient availability are variable.

Figure 5.2 shows the prediction interval of the equation that fit the data presented in Figure 5.1, and the data of KNP and PGR obtained

FIGURE 5.2. Relationship Between Kernel Number per Plant and Plant Growth Rate at a Period Bracketing Silking for Control Plants Growing Without Nutrient or Water Deficiencies (■), or for Plants Growing with Reduced Nitrogen or Water Availability (▲)

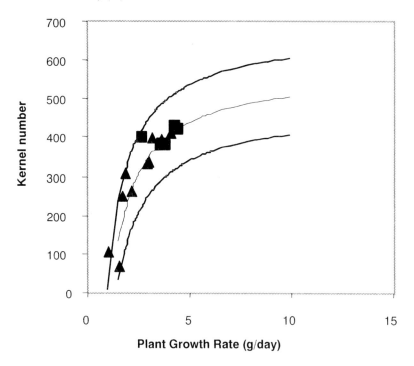

Plant Growth Rate (g/day)

Source: Adapted from Andrade et al. (1999, p. 458).

Note: Hybrid DK636 was sown in October. The general equation fit for data of Figure 5.1 and its 95 percent prediction interval are shown as references. Each point is the average of three replications. Standard errors ranged from 17 to 28 kernels for KNP and from 0.11 to 0.30 g/day for PGR. Plants did not show prolificacy, so kernel number per plant was equal to kernel number per uppermost ear.

from experiments with different levels of nitrogen and water availability. With only one exception, all treatment means values fell within the 95 percent prediction interval of the equation fit for the data in which PGR was varied through plant density or shading (see Figure 5.1). Thus, no evidence of more direct effects of nitrogen or water availability on kernel set were found through the relationship between KNP and PGR. If these more direct effects do exist, they are small and could not be detected by these methods.

The relationship between KNP and PGR obtained for treatments in which PGR was varied through incident radiation per plant could also predict KNP for conditions in which PGR is affected by water and nitrogen deficiencies. Such a model would provide an acceptable prediction of KNP based on PGR under a wide range of environmental conditions.

DURATION OF THE CRITICAL PERIOD FOR KERNEL SET

Dry-matter allocation to reproductive structures between two defined phenological stages is characterized by growth rate and duration of the period (Fischer, 1985). Thus, this second variable must be taken into account to understand kernel set. Duration of the critical period is affected mainly by temperature, and probably by genotype.

When temperature is variable, total dry-matter accumulation in a specific thermal time includes the concept of duration of the CP for grain set. One approach is to calculate total dry-matter accumulation between two defined phenological stages, for example, the window of time for ear elongation, which starts at 227 °Cday before silking and continues up to 100 °Cday after silking (Otegui and Melón, 1997; Otegui and Bonhomme, 1998). However, the critical period for kernel set is not necessarily the period in which the ear elongates (Kiniry and Ritchie, 1985). Kernel abortion could take place beyond 100 °Cday after silking (Cirilo and Andrade, 1994), at the early stages of the grain-filling period.

A similar approach is to calculate plant growth rate per unit thermal time (PGRtt). Plant growth rate per unit thermal time at the period bracketing silking was calculated for the data shown in Figure 5.1, as PGR/t-tb. The base temperature for maize development (tb = 8°C) was subtracted from the average temperature (t) at the period bracketing silking corresponding to each experiment. The relationship between KNP and plant growth rate per unit thermal time was similar to that obtained between KNP and PGR at flowering (see Figure 5.3). This is because temperature differences among treatments were small. Both total growth in a specific thermal period and

FIGURE 5.3. Relationship Between Kernel Number per Uppermost Ear and Plant Growth Rate at a Period Bracketing Silking (A) and Between Kernel Number per Uppermost Ear and Plant Growth Rate per Unit Thermal Time at a Period Bracketing Silking (B)

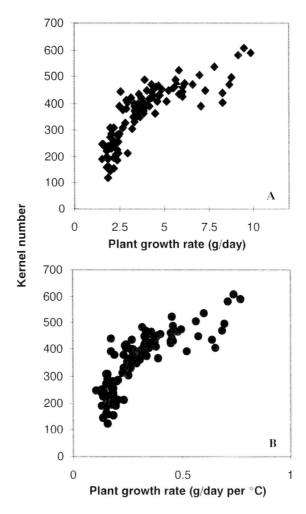

Source: Adapted from Andrade et al. (1999).

Note: Hybrid DK636 was sown in October. PGR changed because of plant density, incident radiation, and year effects. Inverse equations for uppermost ear only: KNP = 573 − 676/PGR (R^2 = 0.78), and KNP = 563 − 50/PGRtt (R^2 = 0.72).

plant growth rate per unit thermal time include the concept of duration of the critical period for grain number determination and are better variables to estimate kernel number when temperature varies among situations. This was the case when night temperature was artificially modified (Andrade et al., 1999). When night temperature was increased, fewer kernels were produced, even though PGR was not affected (Cantarero, Cirilo, and Andrade, 1999). Plant growth rate expressed per unit thermal time was better correlated with KNP.

High night temperatures did not affect PGR, but they did reduce kernel set because of a reduction in the duration of the critical period for kernel number determination. Increased night temperatures hastened crop developmental rate so that the lag period during early kernel growth was reduced by 5 ± 0.36 days. Thus, high night temperatures decreased radiation interception per unit thermal time, reducing the amount of energy available for kernel set. This could have increased competition for assimilates among kernels, magnifying the assimilate shortage for tip kernels and forcing their abortion (Schussler and Westgate, 1991a, b; Boyle, Boyer, and Morgan, 1991). This approach does not eliminate the possible involvement of hormones in modulating sink strength (Morris, 1996).

OTHER FACTORS AFFECTING KERNEL SET

Plant growth per unit thermal time at the period encompassing flowering is not the only factor that influences the number of reproductive sinks set per plant. Partitioning of dry matter to the ear and the number of kernels set per unit of biomass allocated to the ears during the critical period (grain efficiency factor, GEF) must also be considered.

At a specific range of plant growth rates, plants or hybrids could differ in the number of kernels set per unit PGR (Luque et al., 1997; Echarte et al., 1998). For example, the new hybrid DK752 has more kernels per plant at high PGR, more kernels per plant at low PGR, and probably a lower threshold value for kernel set than the older hybrid M400 (see Figure 5.4) (Echarte et al., 1998). At high values of PGR, prolific genotypes, or genotypes with high potential kernel number, set more grains per unit growth (Echarte et al., 1998). The higher number of kernels at low PGR in new hybrids can be explained by a higher

FIGURE 5.4. Relationship Between Kernel Number per Plant (KNP) and Plant Growth Rate (PGR) During the Critical Period for Yield Determination for Two Maize Hybrids, M400 and DK752

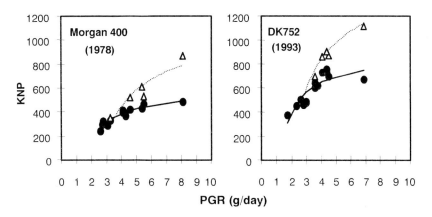

Source: Echarte et al. (1998, p. 102).

Note: Kernels in the uppermost ear (●) and total kernel number in the plant (Δ) are shown.

dry-matter partitioning to the ear at the CP, by a greater GEF, and/or by a longer duration of the CP. These aspects are currently being investigated (see Chapter 11). In an intolerant hybrid at high densities (small PGR), the detasseled or male-sterile plants produced more grains at equal PGR than the male-fertile control plants. An increase in dry-matter partitioning to the ear in detasseled plants would be at least partially responsible for the greater number of kernels per unit plant growth rate compared to control plants (Frugone, 1994).

Dry-matter partitioning to ears is also a function of PGR at flowering and, indirectly, of the environmental factors that affect PGR. Partitioning to the ear decreased at low and high PGR, with maximal values at PGR of 2 to 3 g/day (Andrade et al., 1999). At low PGR, it decreased because the ear is a weak sink within the plant at flowering, probably because it is in an axillary position subjected to apical dominance (Tollenaar, 1977; Gardner and Gardner, 1983). At high PGR, partitioning to the ears decreased because of morphogenetic limitations in the uppermost ear and because of high threshold values for second-ear growth (Ruget, 1989; Edmeades and Daynard, 1979; Andrade et al., 1999). Decreases in PGR below 2 g/day produced strong decreases in ear growth rate, with threshold values for ear growth close to 1 g/plant per day. Moreover, beyond a PGR of 6 g/day, first-ear growth rate showed no response to increases in PGR. For these values of PGR, the plants that showed prolificacy presented higher ear growth rate and higher dry-matter partitioning to the ears (considering the sum of uppermost and second ears).

The grain efficiency factor is a function of the genotype and, probably, of the ear growth rate. Variations in this variable among hybrids can be explained by differences in dry-matter partitioning within the ear, in threshold values of kernel growth for abortion, in dominance effects among kernels, and/or in potential kernel number. One way to increase kernel set at equal ear dry weight is to achieve a more synchronous pollination. A more synchronous ovary pollination would decrease dominant effects among kernels and between ears and would allow more kernel set per unit crop or ear growth (Cárcova, Otegui, and Andrade, 1998).

CONCLUSIONS

A general model of kernel number determination in maize is presented in Figure 5.5. Plant density, radiation, temperature, and nitrogen and water availability affect PGR. Hence, PGR is a good predictor of KNP for a wide range of environmental conditions. Plant growth rate per unit thermal time or total plant growth during a specific critical period encompassing flowering defined in thermal units better explains KNP for conditions in which duration of the critical period varies because of temperature. The genotype can also affect PGR and duration of CP. Plant growth during the CP and the proportion of this growth that is partitioned to the ear, determine total ear growth during this period. This last variable is affected by PGR (and, indirectly, by the environmental factors that affect PGR) and by the genotype. KNP is a function of total ear growth during the CP and of the GEF. This last variable is affected by the genotype and, probably, by plant or ear growth.

FIGURE 5.5. General Model for Kernel Number Determination in Maize

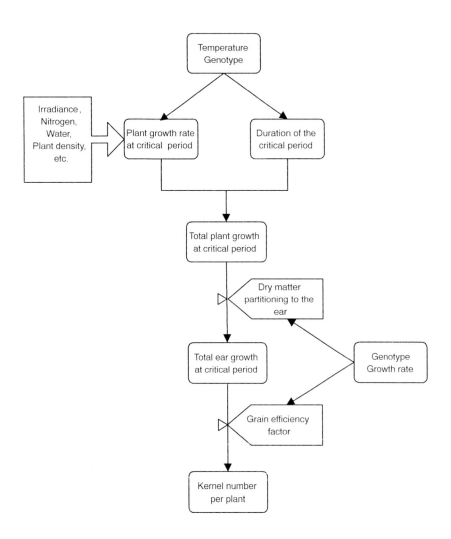

REFERENCES

Aluko, G.K. and Fischer, K.S. (1988). The effect of changes of assimilate supply around flowering on grain sink size and yield of maize (*Zea mays* L.) cultivars of tropical and temperate adaptation. *Australian Journal of Agricultural Research* 39: 153-161.

Andrade, F.H., Cirilo, A.G., Uhart, S.A., and Otegui, M.E. (1996). *Ecofisiología del cultivo de maíz*. Balcarce, Argentina: La Barrosa, CERBAS, and Dekalb Press.

Andrade, F.H., Uhart, S.A., and Cirilo, A.G. (1993). Temperature affects radiation use efficiency in maize. *Field Crops Research* 32: 17-25.

Andrade, F.H., Vega, C.R., Uhart, S.A., Cirilo, A.G., Cantarero, M.G., and Valentinuz, O. (1999). Kernel number determination in maize. *Crop Science* 39: 453-459.

Below, F.E., Cazetta, J.O., and Seebauer, J.R. (1998). Carbon/nitrogen interactions during ear and kernel development. *Annual Meeting Abstracts*, p. 94.

Boyle, M.G., Boyer, J.S., and Morgan, P.W. (1991). Stem infusion of liquid culture medium prevents reproductive failure of maize at low water potential. *Crop Science* 31: 1246-1252.

Cantarero, M.G., Cirilo, A.G., and Andrade, F.H. (1999). Night temperature at silking affects kernel set in maize. *Crop Science* 39: 703-710.

Cárcova, J., Otegui, M.E., and Andrade, F.H. (1998). Polinización sincronizada y determinación del número de granos en maíz. I. Crecimiento de espigas y efecto de la temperatura de espiga. *Actas XXII Reunión Argentina de Fisiología Vegetal* September 23-25 (pp. 118-119). Mar del Plata, Buenos Aires, Argentina: Asociación Argentina de Fisiología Vegetal.

Cirilo, A.G. and Andrade, F.H. (1994). Sowing date and maize productivity. II. Kernel number determination. *Crop Science* 34: 1044-1046.

Claassen, M.M. and Shaw, R.H. (1970). Water deficit effects on corn. II. Grain components. *Agronomy Journal* 64: 652-655.

Czyzewicz, J.R. and Below, F.E. (1992). Growth of maize kernels in vitro as affected by nitrogen supply. *Agronomy Abstracts*, p. 124.

Echarte, L., Vega, C.R., Andrade, F.H., and Uhart, S.A. (1998). Relación entre el número de granos fijados y la tasa de crecimiento por planta en híbridos de maíz de las últimas tres décadas en Argentina. *Actas XXI Reunión Argentina de Fisiología Vegetal* September 23. Mar del Plata, Buenos Aires, Argentina: Asociación Argentina de Fisiología Vegetal.

Edmeades, G.O. and Daynard, T.B. (1979). The relationship between final yield and photosynthesis at flowering in individual maize plants. *Canadian Journal of Plant Science* 59: 585-601.

Fischer, K.S. and Palmer, F.E. (1984). Tropical maize. In Goldsworthy, P.R. and Fisher, N.M. (Eds.), *The physiology of tropical field crops* (pp. 213-248). Chichester, UK: John Wiley and Sons Ltd.

Fischer, R.A. (1985). Number of kernels in wheat crops and the influence of solar radiation and temperature. *Journal of Agricultural Science, Cambridge* 105: 447-461.

Frugone, M. (1994). Efectos del despanojado y la machoesterilidad sobre la tolerancia a la densidad poblacional. MSc thesis. Faculty of Agricultural Sciences, National University of Mar del Plata.

Gardner, W.R. and Gardner, H.R. (1983). Principles of water management under drought conditions. *Agricultural Water Management* 7: 143-155.

Gifford, R.M., Thorne, J.H., Hitz, W.D., and Giaquinta, R.T. (1984). Crop productivity and photoassimilate partitioning. *Science* 225: 801-808.

Jacobs, B.C. and Pearson, C.J. (1991). Potential yield of maize, determined by rates of growth and development of ears. *Field Crops Research* 27: 281-298.

Kiniry, J.R. and Ritchie, J.T. (1985). Shade-sensitive interval of kernel number of maize. *Agronomy Journal* 77: 711-715.

Luque, S.F., Cirilo, A.G., Otegui, M.E., and Andrade, F.H. (1997). Caracteres fisiológicos asociados al mejoramiento de maíz en la Argentina en los últimos 30 años. *Actas VI Congreso Nacional de Maíz*, Volume III (pp. 270-277). November 13-15. Pergamino, Buenos Aires, Argentina: Asociación de Ingeneiros Agrónomos del Norte de la Provincia de Buenos Aires.

Morris, D.A. (1996). Hormonal regulation of source-sink relationships: An overview of potential control mechanisms. In Zamski, E. and Schaffer, A. (Eds.), *Photoassimilate partitioning in plants and crops* (pp. 441-465). New York: Marcel Dekker.

Otegui, M.E. (1997). Kernel set and flower synchrony within the ear of maize. II. Plant population effects. *Crop Science* 37: 448-455.

Otegui, M.E. and Andrade, F.H. (2000). *New relationships between light interception, ear growth and kernel set in maize.* In Mickelson, S. (Ed.), *Physiology and modeling kernel set in maize,* Special publication number 29 (pp. 89-102). Madison, WI: Crop Science Society of America.

Otegui, M.E., Andrade, F.H., and Suero, E.E. (1995). Growth, water use, and kernel abortion of maize subjected to drought at silking. *Field Crops Research* 40: 87-94.

Otegui, M.E. and Bonhomme, R. (1998). Grain yield components in maize. I. Ear growth and kernel set. *Field Crops Research* 56: 247-256.

Otegui, M.E. and Melón, S. (1997). Kernel set and flower synchrony within the ear of maize. I. Sowing date effects. *Crop Science* 37: 441-447.

Prior, C.L. and Russell, W.A. (1975). Yield performance in nonprolific and prolific maize hybrids at six plant densities. *Crop Science* 15: 482-486.

Ruget, F. (1989). Relations entre matière sèche à la floraison et le rendement en grains chez le maïs: Importance du rayonnement disponible par plante. *Agronomie* 9: 457-465.

Schussler, J.R. and Westgate, M.E. (1991a). Maize kernel set at low water potential. I. Sensitivity to reduced assimilates during early kernel growth. *Crop Science* 31: 1189-1195.

Schussler, J.R. and Westgate, M.E. (1991b). Maize kernel set at low water potential. II. Sensitivity to reduced assimilate supply at pollination. *Crop Science* 31: 1196-1203.

Tollenaar, M. (1977). Sink-source relationships during reproductive development in maize: A review. *Maydica* 22: 49-75.

Tollenaar, M. and Daynard, T.B. (1978). Effect of defoliation on kernel development in maize. *Canadian Journal of Plant Science* 58: 207-212.

Tollenaar, M., Dwyer, L.M., and Stewart, D.W. (1992). Ear and kernel formation in maize hybrids representing three decades of grain yield improvement in Ontario. *Crop Science* 32: 432-438.

Uhart, S.A. and Andrade, F.H. (1995). Nitrogen deficiency in maize. I. Effects on crop growth, development, dry matter partitioning, and kernel set. *Crop Science* 35: 1376-1383.

Uhart, S.A., Frugone, M.I., Terzoli, G., and Andrade, F.H. (1995). Androesterilidad en maíz y tolerancia al estrés. *Technical Bulletin No. 136.* Balcarce, Argentina: National Institute of Agricultural Technology, 23 pp.

Vega, C.R. (1997). Número de granos por planta en soja, girasol y maíz en función de las tasas de crecimiento por planta durante el período crítico de determinación del rendimiento. MSc thesis. Faculty of Agricultural Sciences, National University of Mar del Plata, Buenos Aires, Argentina.

Chapter 6

Maize Improvement
for Drought-Limited Environments

Gregory O. Edmeades
Marianne Bänziger
Jean-Marcel Ribaut

INTRODUCTION

It is generally agreed that the two largest causes of maize yield loss in the tropics and subtropics are drought and inadequate fertility, and, for this reason, breeding for tolerance to these stresses has become a major focus of CIMMYT's Maize Program. There is also evidence (E. Knapp, unpublished data; Reeder, 1997) that maize yields in major temperate production areas, such as the central Corn Belt of the United States, even though they have risen substantially in the past fifty years, have shown much greater fluctuations from year to year in the last thirty years. This is attributed mainly to seasonal variation in rainfall, since soil fertility levels have been rising and weed control is generally excellent. Losses to drought alone in the tropics may total 24 million tonnes (t) per year, or around 17 percent of well-watered production (Edmeades, Bolaños, and Lafitte, 1992), but have been as high as 60 percent in severely affected regions such as southern Africa in 1991 to 1992. Severe losses such as these can be expected to lead to a close association between seasonal rainfall and national or regional average maize yields, and this has been observed in eastern and southern Africa (Edmeades, Bänziger, Chapman, Ribaut, and Bolaños, 1997). This suggests that drought is a pervasive cause of yield instability in maize-based cropping systems in most years and environments.

Within a particular field, localized yield losses can reach 100 percent where rainfall falls below 350 to 500 millimeters (mm) in the crop season, or where there are patches of soil that are sandy or shallow, provided the period of severe water deficit coincides with flowering. Eyhérabide and colleagues (1997), in a simulation study of major maize-producing areas of Argentina, noted that in the region where yields are the highest, there is a 50 percent probability of at least a 50 mm deficit in water during the flowering period. They concluded that the average annual yield loss caused by drought in Argentina equaled 1.2 million tons of grain, or about 15 percent of annual production.

The focus of this chapter will primarily be on drought tolerance, but it will be shown that there are spillovers in tolerance from drought to environments where production is limited by nitrogen (N) levels. This chapter concentrates on genetic improvement of drought tolerance because seed-based technology is easier to transfer to small-scale farmers than knowledge-based agronomic practices. We are very much aware, however, that agronomic practices such as timely sowing at the correct density and improved infiltration and reduced evaporation through weed control, mulch management, and tillage (Fischer, 1989; Eyhérabide, Guevara, and Totis de Zeljkovich, 1997) also play an important part in alleviating effects of water shortage in maize.

The following account draws heavily from previously published papers (Edmeades, Bänziger, Chapman, Ribaut, and Elings, 1997; Edmeades et al., 1998; Edmeades et al., 1999; Heisey and Edmeades, 1999).

THE TARGET ENVIRONMENT

Knowledge of the target environments is essential when developing strategies for improving crop tolerance to drought. Maize "mega-environments" can be defined broadly as follows:

1. Summer tropical rainy season: Maize is sown at the onset of the rains. Probabilities of moisture deficit are highest at the start and end of the rains, but dry spells occur unpredictably throughout the season. Rainfall may be unimodal or bimodal.
2. Summer temperate crop season: Maize is sown in the spring when soil temperatures exceed 10 to 14°C and when soils are

usually saturated. Dry spells, often accompanied by high temperatures, occur unpredictably throughout the season, but their effects decrease in the autumn as potential evapotranspiration (ET_0) declines with falling temperatures.

Of much less importance are dry winter production environments in south Asia and Latin America, where maize is usually produced on reserve soil water or with irrigation.

Seedling and terminal stresses are clearly more important in the tropics than in temperate areas, but both maize megaenvironments experience water deficits that are unpredictable in timing and severity during the crop season. Typically, in the more drought prone of these environments, the crop experiences little or no stress in 40 percent of years, a yield reduction of around 25 percent occurs another 40 percent of years, and severe losses, exceeding 50 percent, may occur in the remaining 20 percent. If the frequency of severe stress is greater than this, a more drought-hardy species is usually grown, or irrigation is used. A key feature of such environments is the occurrence of years without any stress. Spatial variation of drought within a single field is another characteristic typical of most farms, especially in the tropics. This is caused by variation in soil texture and depth, in topography, and in infiltration rates. Thus, because the incidence of drought cannot be predicted, and because of spatial variation within a field, it is essential that a drought-tolerant variety be able to perform well under well-watered conditions as well as under drought. Drought tolerance cannot be achieved at the cost of yield under circumstances when crop water supply is adequate.

CONCEPTUAL FRAMEWORK

Grain yield (GY) can be considered the product of the following:

$$GY = RAD \cdot \%RI \cdot GLD \cdot RUE \cdot HI \tag{6.1}$$

where:

RAD = incident solar radiation received per day
      (e.g., 20 megajoules per square meter [$MJ \cdot m^{-2}$])
%RI = percent intercepted radiation over crop life cycle
      (e.g., 50 percent)

GLD   = green leaf duration (e.g., 100 days)
RUE   = radiation use efficiency over the life of the crop, taken as 1.5 grams (g) $\cdot$ MJ^{-1}
HI   = harvest index (0.45) (range 0.4 to 0.55 under well-watered conditions)

Thus:

$$GY = (20 \cdot 0.5 \cdot 120 \cdot 1.5) \cdot 0.45$$
$$= 810 \text{ g} \cdot \text{m}^{-2} \text{ or } 8.1 \text{ tonnes per hectare (t} \cdot \text{ha}^{-1})$$

Passioura (1977) proposed a parallel way of considering grain yield in a water-limited situation:

$$GY = W \cdot WUE \cdot HI \tag{6.2}$$

where:

W   = water transpired by the crop (e.g., 450 mm)
WUE   = water use efficiency, biomass/unit water transpired (e.g., 4 g $\cdot$ m^{-2} $\cdot$ mm^{-1})

Thus:

$$GY = (450 \cdot 4) \cdot 0.45$$
$$= 810 \text{ g} \cdot \text{m}^{-2}, \text{ or } 8.1 \text{ t} \cdot \text{ha}^{-1}$$

Finally, considering yield components:

$$GY = \text{plants/m}^2 \cdot EPP \cdot GPE \cdot WPG \tag{6.3}$$

where:

plants/m^2 = stand established (e.g., 5.5 plants/m^2)
EPP   = ears per plant (e.g., 1.1)
GPE   = grains per ear (e.g., 446)
WPG   = weight per grain (e.g., 0.3 g)

Thus:

$$GY = (5.5 \cdot 1.1 \cdot 446 \cdot 0.3)$$
$$= 810 \text{ g} \cdot \text{m}^{-2} \text{ or } 8.1 \text{ t} \cdot \text{ha}^{-1}$$

Grain yield can be reduced by drought because of direct effects on stand, leaf area, radiation interception (%RI), accelerated senescence (reduced GLD), barrenness (reduced EPP, GPE), and completeness of grain filling (reduced WPG). When water deficit reduces EPP, GPE, and WPG, HI will also fall. Drought may also reduce RUE and possibly increase WUE, though effects on these are generally less important than those on %RI and HI. Impacts of changes in these variables on yield have been discussed fully by Andrade and colleagues (1996).

DROUGHT STRESS AND STAGES OF GROWTH

Seedling Establishment

A first step in obtaining yield is to have sufficient plant stand. If drought severely reduces stand at the onset of the season, farmers can replant fields with a shorter-duration cultivar or a different species, though this entails additional cost. A limited research effort directed toward improving seedling establishment at CIMMYT showed only modest increases in survival under water deficit (Bänziger, Edmeades, and Quarrie, 1997), while selection for decreased survival resulted in significant reductions in capacity to survive under drought. The authors concluded that selection for improved survival and biomass production under postemergence drought stress is difficult because environmental variation is high in field screens, and because natural selection may have exploited most of the genetic variation for these traits.

Flowering

Approximately 59 percent of crop production globally results from organs that are the result of sexual reproduction (Edmeades et al., 1998). Despite its importance, there has been relatively less research on mechanisms determining reproductive success under drought stress than on factors affecting vegetative growth (Saini and Lalonde, 1998). A failure of the rains at flowering, when replanting is no longer possible, may lead to a total loss of grain yield. Monthly rainfall totals in the tropics have a high coefficient of variability, even though mean rainfall may appear adequate for maize production.

Maize grain yield is reduced two to three times more when water deficits coincide with flowering, compared with other growth stages (Shaw, 1977; Grant et al., 1989). Maize is thought to be more susceptible than other rain-fed crops because of its unusual floral structure, notably, the separation of male and female floral organs and the near-synchronous development of florets on a single ear, borne on a single stem.

One indicator of a high growth rate per female spikelet at flowering in maize appears to be rapid silk extrusion. Since time from sowing to anthesis is little affected by drought, rapid silk emergence is reflected in a short anthesis-silking interval (ASI). Plants with a large ASI under drought are often barren or have few grains per ear. Grain yield of maize grown under severe water stress at flowering and during grain filling is highly correlated with kernel number per plant ($r = 0.90$; $P < 0.01$) (Bolaños and Edmeades, 1996). Factors affecting grain set under drought are extensively reviewed by Westgate (see Chapter 7) and are summarized briefly here. Grain number per plant in water-deficient maize appears to depend directly on the flux of current photosynthate during the ten to fourteen days either side of flowering (Schussler and Westgate, 1995), implying that reserves of preflowering assimilate held in the stem are scarce or are simply not attracted to the ear. Furthermore, acid invertase activity in ovaries of water-stressed plants is much reduced (Zinselmeier et al., 1995; Westgate, 1997; Saini and Lalonde, 1998), impairing sink strength because assimilates cannot be converted rapidly to starch.

The critical step in determining HI appears to take place in those ten to fourteen days either side of flowering. During that time, it seems that growth of the ear is also susceptible to competition from other organs that are still growing, and slow ear growth is associated with a long ASI. Calculations based on radiation intercepted by different layers in the canopy, and carbon 14 (^{14}C)-labeled assimilate patterns from leaves located in those layers, show that at 5 plants/m^2 assimilate flows to stem and tassel were 3.9 and 0.5 that of the flux to the cob, whereas at 15 plants/m^2 these had risen to 19.2 and 2.4, respectively (see Table 6.1). Although drought may not result in such

TABLE 6.1. Fluxes of ^{14}C-Labeled Assimilate (g/day) to Various Plant Parts During the First Four Days Following Anthesis in a Short-Season Corn Belt Hybrid Grown at Three Plant Densities in Canada

	Density (plants/m^2)			
Plant part	**5.0**	**10.0**	**15.0**	**Mean**
Stem + sheath	2.10a	1.37b	0.96c	1.48
Tassel	0.27a	0.19b	0.12b	0.19
Cob	0.54a	0.260b	0.11c	0.30
Husk	0.73a	0.15b	0.05c	0.31
Total shoot	4.15a	2.29b	1.47c	2.64
Ratios of fluxes as percentages				
Tassel/shoot	6.5	8.3	8.2	7.2
(Cob+husk)/shoot	30.6	17.9	10.9	23.1
Barrenness (%)	0.0	0.8	9.2	3.3
ASI* (days)	−0.4a	1.2b	3.4c	1.4

Source: Adapted from Edmeades and Daynard (1979).

Note: Means in the same row followed by different letters are significantly different ($P < 0.05$).

* Anthesis-silking interval

large proportional reductions in assimilate flow to the ear, the tendency to maintain tassel growth at the expense of ear growth as drought intensifies has been observed by Bolaños and Edmeades (1993b), and this certainly seems one reason for the delay in silk emergence. Reduced plant height has also been associated with a reduced ASI and an increased tolerance to drought (Fischer, Johnson, and Edmeades, 1983). Selection for smaller tassel size has resulted in increased ear growth at high plant densities (Fischer, Edmeades, and Johnson, 1987). Moreover, although little is known about competing effects of root growth on ear growth, Bolaños, Edmeades, and Martinez (1993) reported a reduction in root biomass and an increase in ear growth under drought in one tropical maize population that underwent selection for increased drought tolerance. In summary, then, it appears that ASI is an easily observed external indicator of partitioning of current assimilate to the ear, of female spikelet growth rate, of grain number, and perhaps also of plant water potential (Edmeades et al., 2000).

What is the role of pollen supply and pollen viability in determining kernel numbers? When silks whose appearance has been substantially delayed because of drought are pollinated with fresh pollen, kernel abortion is frequently observed, even though pollination can be shown to have occurred (e.g., Westgate and Boyer, 1986). Silks, themselves, have a natural life cycle and eventually senesce near their base (Bassetti and Westgate, 1993). On the other hand, Westgate (1997) has shown a dependence of kernel set on pollen shed density, and, thus, pollen supply that falls below 100 grains/ square centimeter (cm^2) per day (or about 20 percent of peak daily production) may limit grain set in a uniform field of hybrid maize. Drought stress itself also reduces the production of pollen (Hall et al., 1982). Pollen viability seems much less likely to be an important factor affecting grain set in populations of maize for which flowering dates are inherently variable (as in open-pollinated varieties), unless temperatures greater than 38°C are encountered at flowering (Schoper, Lambert, and Vasilas, 1986).

Grain Filling

Once kernels enter the linear phase of biomass accumulation, they develop the sink strength needed to remobilize C reserves in the stem and elsewhere. This, and continued photoassimilation, determines final kernel weight. *Provided an ear has been established,* the maintenance of a green functional canopy and a capacity to remobilize carbohydrates stored in the stem and husk should contribute to high yield under terminal stress. Associations between foliar "stay green" and yield are often weak (Bolaños and Edmeades, 1996), and reasons for this must be sought in the N balance of the crop at that growth stage. If more grains are set per plant, the internal demand for N by those additional grains will rise. Since uptake of N from a dry soil is low, this may result in "mining" of N from the leaves, thus offsetting improvements in stay green resulting from directed selection (Chapman and Edmeades, 1999).

EARLINESS, YIELD POTENTIAL, AND THE SELECTION ENVIRONMENT

Earliness versus Tolerance

Where stable grain production is our goal, the first essential step is to match the length of the crop cycle with the duration of water supply in the majority of years, and to synchronize stress-susceptible growth stages with times when the probability of water deficits is lowest. Crop maturity, since it is a highly heritable trait, is easily modified by breeders seeking a better fit between crop and season length. Because loss of yield potential under unstressed conditions inevitably accompanies earliness (reduced GLD, equation 6.1), we advocate using cultivars with a maturity suited to the wettest 20 percent of years in the target environment, but with increased drought tolerance so that improvements in yield potential and stability occur together.

Spillovers

Yield potential is of great importance in drought breeding, since a cultivar must remain competitive in seasons when water is not a limitation. Conventional selection in temperate maize has increased stress tolerance and yield potential over the past seventy years (Castleberry, Crum, and Krull, 1984; Duvick, 1997). These improvements can be traced to the use of self-pollination, high plant densities in breeding nurseries, and multilocation testing. Heterosis, because it contributes to yield potential, must also be considered a source of drought tolerance (Blum, 1997), and hybrids usually outyield varieties in stressed environments (CIMMYT, unpublished data, 1996; see Table 6.4, p. 96). CIMMYT's experience with maize shows that genetic correlations between yield in unstressed and stressed environments remain positive but tend toward nonsignificance where stress reduces yields by around 50 percent. Selection under high N conditions for performance under low N was significantly ($P < 0.05$) more efficient than selecting under low N when yields were reduced less than 40 percent by N stress (Bänziger, Bertrán, and Lafitte, 1997), and it seems likely that a similar situation exists for drought. If the target environment includes a reduction in grain yield of greater than 40 percent compared with unstressed conditions, yield potential should be combined with specific secondary traits that conserve yield under moderate or severe stress and provide improved stability. The greater the yield reduction, the more important drought-adaptive secondary traits are likely to become. If negative genetic correlations exist between yield in unstressed and stressed environments, this would indicate a need to breed for each environment separately.

Multilocation Testing versus Managed-Stress Nurseries

Conventional breeding for drought tolerance has relied on extensive multilocation testing of progenies and has successfully increased grain yields under well-watered and moderately stressed environments (Jensen and Cavalieri, 1983; Castleberry, Crum, and Krull, 1984). Where grain yield is strongly affected by susceptibility at a specific growth stage and rainfall is erratic in distribution, evidence shows that the use of nurseries where timing and intensity of water deficits are carefully managed is more efficient and usually cheaper than multilocation testing. Managed-stress nurseries provide a selection environment that displays genetic variation for drought-adaptive traits to best advantage, even if the stress used for this purpose is more severe than that encountered in the target environment.

THE USE OF DROUGHT-ADAPTIVE SECONDARY TRAITS IN BREEDING

Putative secondary traits indicative of drought tolerance have been reviewed extensively (Blum, 1988; Ludlow and Muchow, 1990; Fukai and Cooper, 1995; Subbarao et al., 1995; Boyer, 1996), though very few of these have proven useful in plant breeding programs. The following discussion draws heavily on the work of Edmeades and colleagues (1998).

Ideally, a secondary trait should be (1) genetically associated with grain yield under drought, (2) highly heritable, (3) genetically variable, (4) cheap and fast to measure, (5) nondestructive, (6) stable over the measurement period, (7) observed at or before flowering, so that undesirable parents are not crossed, (8) an estimator of yield potential before final harvest, and (9) not associated with yield loss under unstressed conditions.

The value of a secondary trait to the selection process can be assessed by analyses of correlation and heritability among progenies of a single population; by divergent selection for that trait or groups of traits to create isopopulations; by changes that have occurred in populations subject to selection for grain yield under specific environmental circumstances; by simulation modeling; or by statistical procedures based on selection index theory. Using this last approach, Bänziger and Lafitte (1997) determined that the use of secondary traits plus yield during selection for tolerance of maize to low soil N was about 20 percent more efficient than selection for yield alone. The authors noted that this benefit increased as yield levels declined. Very few putative secondary traits have passed the tests described previously.

In the following section we focus on indirect traits that integrate the effects of several basic processes, or single traits with demonstrated effects on improving grain production in plant breeding programs. Traits associated with survival (e.g., abscisic acid [ABA] accumulation; Setter, 1997) diminish in value as the crop develops. Thus, during seedling establishment, survival is more valuable than production, whereas productivity-enhancing traits that maintain photosynthate flux to the developing ear become more important during flowering and grain filling.

Osmotic Adjustment (OA)

Provided it is an active rather than a passive process, OA serves primarily to maintain the activity of meristems. Osmotic adjustment may affect grain yield by increasing W through greater water extraction from the soil, increased WUE through maintenance of stomatal conductance (Sinclair et al., 1975), and increased HI through delayed foliar senescence during reproductive development (Sadras and Connor, 1991). The use of OA to identify superior parents in crops such as wheat, rice, and sorghum, whose OA exceeds 1 megapascal (MPa) (Ludlow and Muchow, 1990; Nguyen, Chandra Babu, and Blum, 1997), seems justified. In many crops, however, genetic variation for OA is small (e.g., around 0.4 MPa in maize or cowpea; Bolaños and Edmeades, 1991; Subbarao et al., 1995) and is unlikely to affect grain yield. Chimenti, Cantagallo, and Guevara (1997), on the other hand, found that high OA inbreds, when grown under modest water stress, yielded 89 percent of the well-watered control, compared with 73 percent for the low OA group, and that water uptake was 15 percent greater in the high OA group. High OA capacity has also been associated with stable grain yields of hybrids across environments in Argentina that varied in water availability (Lemcoff, Chimenti, and Davezac, 1998).

Rooting Capacity

Where root length density exceeds 0.6 cm $\cdot$ cm^{-3} there is little evidence that a further increase in rooting intensity will improve water uptake (Ludlow and Muchow, 1990). Moreover, it has been noted by Richards (1992) that chickpea and barley extract similar quantities of soil water, even though their root lengths differ by a factor of 10. Increased root growth comes at a carbon cost to the plant, perhaps at a time when C flux to reproductive organs is already limiting reproductive sink size. Increases in grain yield under drought were associated with reduced root biomass in the upper 50 cm of the root profile in one tropical maize population (Bolaños, Edmeades, and Martinez, 1993), and Richards (1991) reported improved stress tolerance in wheat when roots in the top 30 cm of soil were reduced. Increased rooting *depth* seems fully justified (Boyer, 1996), provided water is present at those depths. Measuring roots is difficult in a breeding program. Several studies have used electrical capacitance measurements for rapid assessment of root dimensions. Although electrical capacitance and total root volume are positively correlated (van Beem, Smith, and Zobel, 1998), the measurement is thought not to be able to assess root distribution. CIMMYT's experiences with divergent selection for this trait are described in the section Roots (p. 103).

Canopy Temperature

Infrared thermometry is a quick and relatively accurate means of detecting differences in W. Canopy temperature has been relatively strongly associated with productivity under drought in maize (Fischer, Johnson, and Edmeades, 1983) and wheat (Blum, 1988), though its heritability in inbred progenies of maize is quite low (Bolaños and Edmeades, 1996).

Leaf Rolling and Leaf Angle

Leaf rolling ratings have been used to visually assess water status. Because leaf rolling reduces radiation interception, most breeders consider it deleterious for production. Genotypic differences have been reported in several species (e.g., Turner et al., 1986), and the trait is highly heritable (Bolaños and Edmeades, 1996). Leaf rolling may, however, also indicate structural characteristics of leaves that bear no relationship to W. Erect leaves are expected to be cooler and to have higher radiation and water use efficiencies under drought than their lax counterparts.

Leaf Senescence

Delayed leaf senescence may indicate access to a larger W. "Stay green" in sorghum is probably controlled by a single major gene (Walulu et al., 1994) and is therefore easily modified. Heterosis also delays leaf death under drought in maize (CIMMYT, unpublished data). Since N cannot be taken up readily from dry soil, a small sink may also be the cause of apparent stay green. Chapman and Edmeades (1999) estimate that an increase in grain yield under drought from 2.0 to 2.8 t $\cdot$ ha^{-1} will increase N demand by developing kernels by an amount equal to N contained in 30 percent of the leaf biomass under drought. Evidence also suggests that leaves in maize exhibiting stay green may not always remain metabolically active (Smart et al., 1995; Hauck et al., 1997).

Partitioning and Remobilization

The effects of competing growth of other organs on ear growth at flowering was addressed in the section Flowering (p. 79). Reducing the number of florets growing simultaneously may also improve the chances of reproductive success of those which remain. Recurrent selection over eight cycles lowered spikelet number on the upper ear of tropical maize by 21 percent and markedly improved seed set under drought (Edmeades et al., 1993).

Remobilization of previously fixed C helps maintain a reasonably constant rate of grain filling under stress. Ability to remobilize previously fixed C varies among genotypes and has been identified in wheat by applying chemical defoliants midway through grain filling (Blum, 1998). Contrary to findings by Blum for wheat, taller maize varieties have not proven to be more stable in their yield under drought or grain-filling stress than their shorter counterparts (Fischer, Johnson, and Edmeades, 1983; Edmeades and Lafitte, 1993).

Studies in Maize Progeny Trials

Data from trials of more than 3,500 S_1 progenies from several populations suggest that heritabilities of leaf extension rate, canopy temperature, and leaf chlorophyll concentration are low, and that those for leaf erectness, tassel size, senescence scores, and ASI are relatively high (Bolaños and Edmeades, 1996) (see Table 6.2). Of these, only ASI and tassel size appear to have significant adaptive value under drought, though delayed senescence during grain filling, cool canopy temperatures, and upright leaves also form part of our stress-tolerant water-efficient ideotype designed to grow in a competitive situation. Genotypic correlations between secondary traits and grain yield under drought stress for S_1 families (see Table 6.2) show that variation in yield under stress is almost completely determined by variation in kernel number per plant. The relationship between means for grain yield and ASI, ears/plant, weight/kernel, and kernels/ear for fifty trials of S_1, S_2, or S_3 progenies are shown in Figure 6.1, and Edmeades, Bolaños, and Chapman (1997) provide a general discussion of the relative values of these traits in a practical selection program.

At low yield levels (< 20 percent of yield potential), heritabilities for yield begin to fall, though they are quite stable up till that point (Bolaños and Edmeades, 1996). The genetic correlation between grain yield under severe stress and ASI was -0.60, and that for grain yield and ears per plant 0.90, indicating that these traits become good surrogates for grain yield in these severely stressed environments (see Table 6.2).

FIGURE 6.1. Relationship of Grain Yield (GY) to Anthesis-Silking Interval (ASI), Ears per Plant (EPP), Weight per Kernel (KW), and Kernel Number per Ear (KNE) of Maize Progenies (S_1 to S_3) Grown Under a Range of Available Water

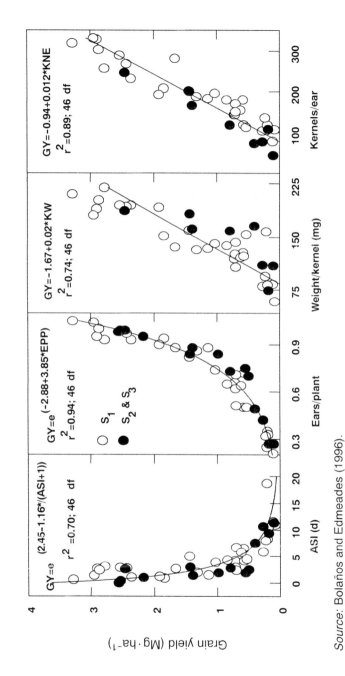

Source: Bolaños and Edmeades (1996).

Note: Data are means of fifty trials containing subsets of a total of 3,509 progenies.

TABLE 6.2. Broad-Sense Heritabilities Observed Under Severe Drought Stress and Genetic Correlations Between Grain Yield and Selected Traits Under Severe Drought Stress for S_1 Progenies Drawn from Several Maize Populations

	No. trials	Heritability under stress	Genotypic correlation
Ears/plant	9	0.54 ± 0.08	0.90 ± 0.14
Kernels/ear	8	0.39 ± 0.13	0.71 ± 0.22
Kernels/plant	8	0.47 ± 0.08	0.86 ± 0.15
Kernel/weight	9	0.43 ± 0.14	0.14 ± 0.17
Days to anthesis	9	0.72 ± 0.08	-0.58 ± 0.12
ASI	8	0.51 ± 0.12	-0.60 ± 0.24
Leaf rolling score	9	0.52 ± 0.09	-0.03 ± 0.15
Leaf erectness score	1	0.74 ± 0.07[†]	-0.28 ± 0.19[‡]
Leaf senescence score	9	0.54 ± 0.08	0.14 ± 0.15
Canopy temperature	4	0.25 ± 0.05	-0.20 ± 0.15
Tassel branch number	1	0.82 ± 0.04[†]	0.15[‡]

Source: For details, see Bolaños and Edmeades (1996).

Note: Heritability of grain yield under severe stress was 0.43 ± 0.10, and yields averaged 14 percent of those of well-watered plots.
  † Trait observed under well-watered conditions.
  ‡ Observed in S_2 or S_3 progenies under severe drought stress.

Selection using ASI and ears per plant, both determined near flowering, can add precision to the routine determination of the yield potential of a genotype under stress. Observations of ASI in a water-stressed nursery can also be used to eliminate drought-susceptible lines from a well-watered breeding nursery sown a few weeks later, without losing a crop cycle to testing under stress.

SELECTION IN ELITE TROPICAL GERMPLASM FOR TOLERANCE TO DROUGHT DURING FLOWERING AND GRAIN FILLING

Detailed accounts of this have been published elsewhere (Bolaños and Edmeades, 1993a, b; Bolaños, Edmeades, and Martinez, 1993;

Byrne et al., 1995; Beck et al., 1996; Edmeades et al., 1993; Edmeades, Bolaños, et al., 1997; Edmeades et al., 1999), and only a summary of procedures and results will be given here.

Selection Procedures

Selection began in the elite lowland tropical white dent population, Tuxpeño Sequía, in 1975. This population underwent eight cycles of recurrent selection among full-sib families in the rain-free winter crop season at Tlaltizapán, Mexico, where timing and intensity of stress can be managed by irrigation. Each of 250 families was grown in single-row plots under three regimes of increasing drought intensity. These were well-watered (WW), intermediate stress (IS; water withdrawn during late flowering and throughout grain filling), and severe stress (SS; no water applied from three weeks before silking onward). Selection of the best sixty to eighty families was based on an ideotype with high grain yield under SS and IS, high leaf and stem extension rates under drought, delayed foliar senescence under SS and IS, and reduced canopy temperatures and ASI under IS and SS. Each cycle of full-sib selection required one year to complete.

Because initial results from selection in this population were encouraging, improvement commenced in a new and diverse group of five elite maize populations in 1985 to 1986, using a recurrent S_1 selection scheme, and continues in three of these (Edmeades, Bolaños, et al., 1997). The population TS6 was derived from Tuxpeño Sequía C_6 but was renamed as TS6 because the progeny structure during selection changed from full-sibs to S_1 families. Three of the five populations (La Posta Sequía, Pool 26 Sequía, and TS6) are late maturing, and two (Pool 16, Pool 18 Sequía) are early maturing. Within each selection cycle, 500 to 600 S_1 families were prescreened under drought and heat in the Sonoran Desert at Obregón, northwest (NW) Mexico. The superior 200 to 220 families thus identified were grown in Tlaltizapán during winter from remnant seed under the three water regimes described, and the best fifty S_1 families were recombined to form the subsequent cycle of selection. The stress level tolerated by S_1 families is less than that for full-sib families, so water was applied once after flowering in the SS treatment. Each of the managed selection environments was used to expose genetic variation for a specific trait. The dry heat of Obregón, where temperatures often exceed 37°C, exposed variation for upper leaf senescence, tassel sterility, and barrenness. At Tlaltizapán, the WW environment allowed expression of yield potential, while the IS regime exposed genetic variation for lower leaf senescence and for grain yield. Under the SS regime, variation in grain yield was small, but genetic variability for ears per plant (barrenness) and for ASI was large. Selection was principally for increased shelled grain yield, ears per plant, and stay green under stress, and for decreased ASI and tassel sterility; less important traits were upright unrolled leaves, small tassels, and lodging resistance. Two years were needed to complete a cycle of two-stage recurrent S_1 selection.

Evaluation of Progress

Evaluations of C_1, C_2, C_4, C_6, and C_8 of Tuxpeño Sequía were conducted in two seasons from 1987 to 1989 in large plots at the selection site, using the three water regimes described. Irrigation regimes were grouped over years into droughted and WW environments. Selection resulted in a significant increase in grain yield of 108 kilograms (kg)$\cdot$ha^{-1} per cycle at yield levels ranging from 1 to 8 t$\cdot$ha^{-1}, and the same rate of gain was observed under WW and drought-stressed conditions (see Table 6.3). At a yield level of 2 t$\cdot$ha^{-1}, this represented a gain of 6.3 percent per cycle (Bolaños and Edmeades, 1993a). These same genotypes, and some conventionally selected checks, were grown in twelve environments, seven of which were rain fed and outside Mexico. Gains averaged 90 kg$\cdot$ha^{-1} per cycle across all sites at a mean yield level of 5.6 t$\cdot$ha^{-1} (Byrne et al., 1995). This indicated that 83 percent of the gains from selection in a dry winter environment were observed in a more normal production situation in the summer rainy season.

In populations subjected to S_1 recurrent selection, gains were evaluated after one to three selection cycles in 1992 to 1994. Gains per cycle were expected to be larger than those under full-sib selection because selection intensity was greater (35 percent of families selected in full-sibs versus 8 percent in S_1). Cycles of selection of La Posta Sequía (C_0, C_1, C_2, and C_3) and Pool 26 Sequía (C_1, C_2, and C_3) were compared with the conventionally improved versions of these populations (Pop. 43 C_9 and Pool 23 C_{23}) and with Tuxpeño

Sequía C_0 and C_8 and TS6 C_1, along with some check cultivars (Edmeades et al., 1999). Average yields of the environments varied from 1.01 to 10.40 t·ha^{-1}. Five environments were droughted during flowering and grain filling, and the other five were well watered. Mean yields were 2.35 t·ha^{-1} and 8.0 t·ha^{-1}, respectively. Yield gains across the two populations, La Posta Sequía and Pool 26 Sequía, averaged 259 kg·ha^{-1} (12.4 percent) per cycle under drought, and 115 kg·ha^{-1} (1.5 percent) per cycle under well-watered conditions (see Table 6.3). Yield improvements under drought were paralleled by increases in ears per plant of around 0.075 per cycle and in harvest index, while ASI declined by 1.3 days per cycle. Despite our objective of maintaining constant time from planting to anthesis, populations became earlier to flower by 0.7 days per cycle. Conventional selection, using international progeny evaluation at mainly well-watered sites, did not increase drought tolerance; Pop. 43 C_9

TABLE 6.3. Effects of Selection for Drought Tolerance on Gains Per Selection Cycle in Four Maize Populations When Evaluated at Three to Six Water-Stressed (SS) Sites, at Five to Eight Well-Watered (WW) Sites, or at Two Low N Sites.

	Grain yield			Anthesis WW	ASI SS	Ears per plant^{-1} SS
Population	SS	WW	Low N			
	kg·ha^{-1}			days		
Evaluation 1988-1989						
Tuxpeño Seq. (Mex.)	100**	125**		-0.40**	-3.0**	-0.03**
Tuxpeño Seq. (Int.)	52ns	101**		-0.24**		
Evaluation 1992-1994						
La Posta Seq. (Mex.)	229**	53ns	187	-0.52**	-1.18**	0.07**
Pool 26 Seq. (Mex.)	288**	177**	199	-0.93**	-1.50**	0.08**
Tuxpeño Seq. (Mex.)	80**	38**	81	-0.32**	-0.44*	0.02**
Pool 18 Seq. (Mex.)	146**	126**	143		-2.13**	0.05**

Source: Beck et al. (1996, p. 91).

Note: Locations were in Mexico (Mex.) or outside (Int.). *, **, ns: significant rate of change per selection cycle at $P < 0.01$, $P < 0.05$, or $P > 0.05$.

and Pool 26 C_{23} became later by about 1 day per cycle, but improvements in yield, if they occurred at all, were only observed under WW conditions. In this evaluation, gains of Tuxpeño Sequía C_8 versus C_0 were slightly lower than those reported previously (Bolaños and Edmeades, 1993a; Byrne et al., 1995), though TS6 C_1 represented an important improvement over Tuxpeño Sequía C_8 under all environments (see Table 6.3). Principal component analysis of yields in the ten different environments showed that well-watered and droughted environments were generally independent of each other (Chapman, Crossa, and Edmeades, 1997), interpreted by the authors as meaning that selection in only WW environments is unlikely to give improvements in yield under drought. They concluded from these, and other analyses, that selection for drought tolerance has improved broad adaptation, as well as specific adaptation to dry environments.

In a similar set of trials to those described previously, rates of progress were determined in the early maturing population, Pool 18 Sequía (C_0 to C_3) (see Table 6.3). Gains averaged 146 $kg \cdot ha^{-1}$ (10.3 percent) per cycle under drought, 126 $kg \cdot ha^{-1}$ (2.4 percent) per cycle under well-watered conditions, and 134 $kg \cdot ha^{-1}$ (3.8 percent) per cycle across sites (see Table 6.3). This contrasted with a slight loss in yield (-3.2 percent) under drought and a 2.4 percent per cycle yield increase under WW conditions in the check entry, Pop. 31, improved by conventional selection based on international progeny testing.

Initial and final selection cycles of the populations were also grown under low and high N over two seasons during 1992 to 1994. Yields under low N averaged a little less than half those under high N, thus constituting a moderate level of N stress. Gains under low N were surprisingly similar to those observed under drought, suggesting that, under moderate levels of N stress, similar mechanisms were enhancing reproductive success. This is partially confirmed by the observed reduction in spikelet number on the upper ear in a population selected for improved grain yield under low N (Lafitte and Edmeades, 1995), a change also reported during selection for drought tolerance (Edmeades et al., 1993).

Recent evaluations of progress during 1997 to 1998 have included the first five cycles of selection in La Posta Sequía, four cycles of selection in TS6, as well as the same three selection cycles of Pool 26

Sequía. This evaluation compared their performance with CIM-MYT's best stress-tolerant hybrids, and was conducted in five severely droughted environments (1.34 t·ha^{-1}, or about 20 percent of potential), three very low N environments (mean yield 1.82 t·ha^{-1}, or < 35 percent of potential), and two unstressed environments (6.38 t·ha^{-1}). Results are shown in Table 6.4. Gains under drought in La Posta Sequía, Pool 26 Sequía, and Tuxpeño Sequía were 154, 163, and 104 kg·ha^{-1} per cycle, respectively, or a little less than that observed in the earlier evaluations. This was probably because the mean yield of the trials was lower than that in other evaluations. Gains under low N averaged 43 percent of those observed under drought. This suggests that, under severe N stress, the spillover of gains under drought into low N environments is not as complete as that observed by Bänziger, Edmeades, and Lafitte (1999), where yields under low N averaged around 2.9 t·ha^{-1}. In all cases, ASI and barrenness were reduced by selection. It is clear from these data that stress-tolerant hybrids will generally outyield open-pollinated varieties under a wide array of conditions, and heterosis must therefore be considered a mechanism for stress tolerance.

Other estimates of gain are provided by trials of original and most advanced cycles of selection for all populations under recurrent selection, repeated under unstressed conditions (three sites), low N (one site), and drought (one site) during the past twelve months. Yields and gains from these trials are presented in Table 6.5, and they generally confirm that drought is a highly effective selection environment for improving performance under low N and under unstressed conditions. It results in greater gains than where the selection emphasis is placed simply on performance under low and high N (Across 8328 BN) or on prolificacy (more than one ear per plant) (semiprolific late [SPL], semiprolific early [SPE], and semiprolific midaltitude tropical [SPMAT]), rather than on resistance to barrenness. Again, hybrid checks outyielded the open-pollinated varieties by 10 to 15 percent across sites, though they were not superior under drought (data not shown).

To sample a wider array of alleles, we screened over 200 landraces and around 100 cultivars with putative drought-adaptive characteristics and composited the best of these to form drought-tolerant populations (DTP) of white and yellow grain color. The adaptation of the

TABLE 6.4. Gains from Selection for Drought Tolerance in Maize Populations La Posta Sequía, Pool 26 Sequía, and Tuxpeño Sequía, when Evaluated Under Drought, Under Low N, and Under Unstressed Conditions, Mexico, 1997 to 1998

Entry	Grain yield				AD	ASI	EPP	Biomass	HI
	Drought	Low N	Unstressed	All envs					
	$t \cdot ha^{-1}$				days			$t \cdot ha^{-1}$	
1. La Posta Sequía C_0	0.91	1.31	5.91	2.03	80.1	8.07	0.67	8.31	0.23
2. La Posta Sequía C_1	0.90	1.52	6.36	2.18	80.5	7.59	0.64	8.73	0.22
3. La Posta Sequía C_2	1.20	1.80	6.66	2.47	80.9	5.36	0.74	9.14	0.25
4. La Posta Sequía C_3	1.41	1.94	6.24	2.53	80.7	4.73	0.76	9.25	0.27
5. La Posta Sequía C_4	1.57	1.93	6.44	2.65	81.6	4.10	0.75	9.35	0.27
6. La Posta Sequía C_5	1.54	1.97	6.71	2.70	81.7	3.21	0.79	9.29	0.28
7. Pool 26 Sequía C_0	0.95	1.46	6.04	2.12	76.4	9.92	0.64	8.32	0.23
8. Pool 26 Sequía C_1	1.14	1.66	5.73	2.21	75.4	7.54	0.69	8.32	0.25
9. Pool 26 Sequía C_2	1.29	1.54	6.04	2.21	75.6	5.81	0.74	8.26	0.26
10. Pool 26 Sequía C_3	1.44	1.59	6.41	2.48	75.3	4.38	0.77	8.29	0.29
11. Tuxpeño Sequía C_0	0.90	1.87	5.70	2.15	80.9	7.91	0.67	8.57	0.24
12. Tuxpeño Sequía C_6	1.56	2.09	6.41	2.69	77.7	3.50	0.74	8.89	0.29
13. TS6 C_4	1.93	2.06	6.33	2.85	77.5	2.21	0.86	8.65	0.33
14. CML258 × CML341	1.26	2.85	7.16	2.92	82.8	4.20	0.70	10.68	0.25
15. CML348 × CML287	1.68	1.77	6.69	2.71	82.4	3.02	0.86	9.41	0.26
16. CML339 × CML344	1.60	1.72	7.30	2.77	84.0	4.30	0.72	10.08	0.26
Mean	1.33	1.82	6.38	2.49	79.6	5.37	0.73	8.97	0.26
LSD (0.05)				0.36	1.8	2.21	0.10	0.82	0.03
Number of environments	5	3	2	10	10	10	10	9	9
Gains per cycle	$kg \cdot ha^{-1}$				d	d	no.	$t \cdot ha^{-1}$	g/g
La Posta Sequía	154**	134*	110ns	138**	0.33**	−1.01***	0.027*	0.196**	0.012**
Pool 26 Sequía	163**	27ns	142*	119*	−0.30ns	−1.84**	0.044**	−0.015ns	0.019*
Tuxpeño Sequía	104*	21ns	67ns	72ns	−0.35ns	−0.58ns	0.018ns	0.012ns	0.009*

Note: ***, **, * signifies gains different from zero at $P < 0.001$, < 0.01, and < 0.05, respectively. AD: days to 50 percent anthesis; ASI: anthesis-silking interval; EPP: ears per plant; biomass: total shoot biomass; HI: harvest index; ns = not significant.

TABLE 6.5. Gains per Selection Cycle Under Drought, Low Nitrogen, and Under Unstressed Conditions in a Series of Populations Undergoing Recurrent Selection, Mexico, 1997 to 1998

		Grain yield								Sene-	Roll
	Drought	Low N	Unstr.	All envs.	AD	ASI	EPP	Biomass	HI	scence	score
		kg·ha^{-1}			days			t·ha^{-1}		(0-10)	(1-5)
Germplasm											
Drt.-Tolerant Pop. (8 cycles)	45	73	140	108	0.19	-0.21	0.018	0.133	0.005	-0.07	-0.06
Across 8328 (7 cycles)	12	0	-38	-23	-0.03	-0.21	0.002	-0.096	0.003	-0.01	0.03
Semiprolific Late (8 cycles)	43	-51	49	28	0.30	-0.34	0.050	-0.060	0.004	-0.05	-0.10
Semiprolific Early (8 cycles)	-28	30	129	78	0.61	-0.04	0.044	0.105	0.002	-0.10	0.14
Semiprolific MAT (7 cycles)	4	0	59	36	0.60	-0.28	0.031	0.077	0.000	-0.15	0.02
La Posta Sequía (6 cycles)	15	63	163	113	0.18	-0.14	0.019	0.039	0.008	-0.14	0.00
Tuxpeño Sequía (10 cycles)	72	27	87	72	-0.32	-0.55	0.020	0.009	0.008	-0.01	-0.01
Pool 16 BNSEQ (4 cycles)	25	90	178	128	-0.03	-0.22	0.010	0.250	0.004	-0.13	0.00
Pool 18 Sequía (5 cycles)	54	106	164	130	0.76	-0.33	0.027	0.212	0.007	-0.16	0.04
Mean trait value	410	1,570	6,570	4,340	72.10	3.13	0.970	10.993	0.338	4.06	2.78
Number of sites	1	1	3	5	5	5	5	5	5	3	1

Note: AD: anthesis date; ASI: anthesis-silking interval; EPP: ears per plant; biomass = total shoot biomass; HI: harvest index; senescence is a visual score of leaf senescence where 0 = no leaf area dead and 10, all leaf area is dead; roll score is a visual score of leaf rolling, where 1 = no rolling, 5 = fully rolled.

97

components of these source populations is approximately 20 percent temperate, 20 percent subtropical, and 60 percent lowland tropical. About 15 percent of the components making up the populations were landraces, mainly from Mexico, and others came from the United States, South Africa, Zimbabwe, Kenya, Thailand, and Nigeria. In other breeding programs, landraces have been a useful source of unique drought-adaptive traits, such as the *latente* syndrome (e.g., Castleberry and Lerette, 1979). However, although the DTPs perform competitively over the full range of water availability, their agronomic defects, especially disease susceptibility and husk cover, remain constraints to their use. In general, we believe that where resources are limiting, national programs should focus on elite germplasm rather than searching for unique drought-adaptive traits in exotic or unimproved germplasm sources.

As a test of broad adaptation, the most advanced cycles of selection of these populations, other drought-tolerant source populations such as DTP1 and DTP2, and conventional and local check entries were included in international cultivar tests as part of CIMMYT's Drought Network activities (see Table 6.6). Late entries were part of a twenty-entry trial grown at nineteen sites, and the early entries were part of a twelve-entry trial grown at twenty-one sites. La Posta Sequía C_3 showed high yields at sites with adequate moisture, as did the conventionally bred elite check, Across 8627 RE. Pool 26 Sequía C_3, a relatively early cultivar, was not stable across sites, performing well in severely stressed sites but poorly in well-watered sites. La Posta Sequía C_3 and several site-specific selections of DTP were identified as being high yielding and stable across sites. Grain yield showed a strong dependence on ears per plant ($r = 0.71**$) and on ASI ($r = -0.51**$). Ears per plant and ASI were also closely associated ($r = -0.55**$). In the early maturing trial (see Table 6.6), a selection from DTP performed well, especially under well-watered conditions. In severely droughted sites, earliness had a clear advantage, and Pool 18 Sequía C_3 was the best performer, and it and Pool 16 C_{20} Synthetic (Syn.) 1 were stable and high yielding across sites. Grain yield over sites was again heavily dependent on ears per plant ($r = 0.68**$), though not on ASI ($r = -0.34$ns). In the severely stressed locations, ears per plant and ASI showed a strong interdependence ($r = -0.83**$) (Edmeades, Bänziger, Chapman, Ribaut, and Bolaños, 1997). Clearly, early maturing culti-

TABLE 6.6. Anthesis Date and Grain Yield (Rank in Parentheses) of Drought-Tolerant Selections, Conventional Checks (RE), and Local Checks Included in Trials of (A) Twenty Late-Maturing Cultivars and (B) Twelve Early Maturing Cultivars, When Grown Internationally, 1991 to 1994

	Sowing to anthesis	Yield, all sites	Yield, severe stress	Yield, moderate stress	Yield, little stress
	days	$t \cdot ha^{-1}$			
A: Late maturing trial					
Number of sites	19	19	5	7	7
La Posta Seq. C_3	64.5	3.58 (1)	0.99 (12)	2.95 (6)	6.06 (1)
Pool 26 Seq. C_3	58.4	2.93 (20)	1.37 (1)	2.36 (20)	4.61 (20)
Ngabu (1) DTP1 C_5	61.6	3.36 (4)	1.08 (6)	2.86 (10)	5.48 (9)
DTP2 C_2	60.5	3.35 (6)	1.08 (8)	2.75 (15)	5.57 (4)
TS6 C_1	64.2	3.31 (9)	0.58 (20)	2.99 (4)	5.57 (3)
Across 8627 (RE)	62.8	3.42 (3)	1.17 (2)	2.91 (8)	5.53 (7)
Local check 2	63.6	3.56 (2)	0.88 (18)	3.36 (1)	5.68 (2)
Mean of trial	61.8	3.30	1.01	2.84	5.39
$LSD_{0.05}$	1.36	0.32	0.39	0.55	0.58
B: Early maturing trial					
Number of sites	21	21	7	7	6
DTP1 C_5 early selection	59.4	2.63 (1)	0.89 (6)	2.70 (1)	4.63 (1)
Pool 18 Sequia C_3	54.3	2.53 (2)	0.99 (3)	2.67 (2)	4.10 (6)
Pool 16 C_{20} Syn. 1	54.2	2.47 (4)	1.03 (1)	2.58 (4)	4.05 (9)
TIWD Drt.-Tol. Pop. C_0	59.7	2.41 (6)	0.67 (12)	2.48 (5)	4.31 (2)
Local check 1	58.7	2.29 (9)	1.00 (2)	1.90 (10)	4.18 (4)
Santa Rosa 8330 (RE)	58.3	2.50 (3)	0.94 (5)	2.38 (8)	4.18 (3)
Mean of trial	56.8	2.38	0.87	2.36	4.11
$LSD_{0.05}$	1.80	0.28	0.40	0.48	0.51

Source: Adapted from Edmeades, Bänziger, Chapman, Ribaut, and Bolaños (1997).

vars possessing a high level of drought tolerance have an important role to play in many national programs.

Finally, a comparison of adapted commercial maize hybrids with hybrids developed from stress-tolerant inbred lines extracted from CIMMYT's drought-tolerant populations, or identified during screening under drought in Mexico, has been conducted recently in southern Zimbabwe under three water regimes (M. Bänziger, personal communication, 1998). Results from trials of 216 hybrids (see Table 6.7) clearly indicate that commercial hybrids developed under mild or no

Table 6.7. Results of a Comparison of Commercial Check with Hybrids Developed from Stress-Tolerant Inbreds Derived from Drought-Tolerant Populations or from Screening of Lines Under Drought, When Evaluated Under Drought in Zimbabwe, 1997

Entry	Sowing to anthesis	Grain yield (rank)		
		No drought[†]	Mild drought	Severe drought
	days	t·ha^{-1}		
(M37W/ZM607#bF37sr//				
P43C$_9$-1)	79.5	12.61 (1)	7.11 (54)	5.47 (1)
SC601	73.9	5.59 (142)	7.16 (52)	2.95 (38)
SC621	74.7	8.61 (48)	9.65 (2)	2.43 (73)
SC623	75.2	9.23 (36)	8.49 (11)	0.84 (206)
SC701	84.6	6.74 (97)	7.44 (41)	0.31 (215)
SC707	78.0	4.22 (186)	5.42 (133)	1.05 (199)
PAN695	74.8	5.70 (134)	6.92 (62)	2.20 (94)
CX5019	77.8	7.53 (70)	6.45 (79)	2.86 (44)
CML202/CML206	83.2	7.39 (76)	6.75 (66)	0.0 (216)
Mean	79.4	6.67	5.98	2.16
LSD$_{0.05}$	2.7	2.98	3.40	1.59

Source: M. Bänziger, personal communication (1998).

Note: Ranks out of 216 entries are in parentheses.

[†] Some infection by gray leaf spot that reduced leaf area affected yield in susceptible entries.

stress are often not suited to conditions of severe drought at flowering, even in their zone of adaptation.

What Changed with Recurrent Selection?

Gains were largely the result of reduced barrenness (i.e., increased ears per plant) under drought and an associated increase in harvest index, since total biomass production was unaffected by selection (Bolaños and Edmeades, 1993a; Edmeades et al., 1999). In one population, La Posta Sequía, gains in biomass of 209 kg · ha^{-1} per cycle were observed over five cycles of selection (see Table 6.4). Tassel size was reduced in Tuxpeño Sequía, and root biomass in the top 50 cm of soil declined by 35 percent; there was no change in any trait indicative of plant water status (e.g., predawn or noon water potential, osmotic adjustment, canopy temperature, water extraction profiles) (Bolaños, Edmeades, and Martinez, 1993). In Tuxpeño Sequía, anthesis date became slightly earlier, and, in all populations, ASI became shorter under drought (see Table 6.4). Eight cycles of selection in Tuxpeño Sequía led to significantly faster spikelet and ear growth, but also to a 21 percent reduction in final spikelet number (Edmeades et al., 1993). Thus, fewer spikelets were formed, grew more rapidly, and were ultimately more successful in forming grain, especially under conditions of drought at flowering. Preliminary data suggest that these changes in ear growth have also occurred in the other late-maturing populations under improvement.

It appears that ASI is a reflection of partitioning to the ear under drought *and* under low N. Success in selection was partly attributed to use of an index of traits that collectively described a drought-tolerant ideotype. Gains under water deficits *were at no cost to yield in unstressed environments,* in part, we believe, because partitioning to the ear and harvest index were increased in all environments. Partitioning requires a carefully managed drought stress to expose the symptoms of its genetic variation (ASI and barrenness) in these diverse maize populations.

RECENT TRENDS

The Relationship Between Stress Tolerance of Populations and Hybrids

Although the research reported here has focused on open-pollinated varieties (OPVs), evidence suggests that these improvements in source populations will carry over to lines and hybrids derived from them. In a recently completed study, 90 to 100 S_2 lines, derived at random from each of three populations improved specifically for drought tolerance or by conventional selection, were crossed to two elite testers (inbred lines), and the resulting hybrids were evaluated under drought. Hybrids derived from the drought-tolerant populations outyielded those derived from their conventionally selected equivalents by 20 percent at a mean yield level of 1.6 t·ha^{-1}. The probability of obtaining a hybrid with a yield 30 to 50 percent greater than this mean value was three to five times greater when lines were derived from drought-tolerant sources, compared with conventionally selected source populations (Edmeades, Bänziger, Cortes, and Ortega, 1997). Population improvement, therefore, has a value in increasing the chance of finding a hybrid with exceptional drought tolerance, and investments in recurrent selection can therefore have a direct payoff in a hybrid breeding program.

Mapping the Traits

Quantitative trait loci (QTL) identification for ASI and yield components in maize has been completed using an F_2 population derived from a cross of lines, P_1 (short ASI) and P_2 (long ASI). For ASI, six putative QTL were identified under drought in this cross. They were located on chromosomes 1, 2, 5, 6, 8, and 10 and accounted for 47 percent of the phenotypic variance. The four QTL positions whose alleles were contributed by the resistant line were responsible for a seven-day reduction of ASI, represented 9 percent of the linkage map, and were stable over years and stress levels (Ribaut et al., 1996; Ribaut and Jiang, 1997). In contrast, all but two yield QTL were inconsistent in genomic position across water regimes. At one important genomic position, the allele contributing to a reduction in ASI also contributed to a grain yield decrease. Because of this, a successful marker-assisted selection strategy should be based on an index of best QTL for both traits.

Two studies, based on marker-assisted selection and the map information previously developed, have been undertaken in CIMMYT. The first aims to improve drought tolerance in the elite but susceptible inbred line CML 247 by transferring favorable QTL from the donor line P_1. At present, topcrosses of BC_2F_3 lines selected with markers for short ASI are being evaluated in the field. The advent of polymerase chain reaction (PCR)-based markers, such as simple sequence repeats (SSRs), has significantly reduced the cost of marker-aided backcrossing and made it possible to screen large numbers (e.g., 2,000 to 3,000) of genotypes at each backcross generation. This significantly shortens the backcrossing procedure. CIMMYT researchers have also attempted to improve drought tolerance in the open-pollinated population Tuxpeño Sequía, using markers to select those alleles which were observed to change in frequency during recurrent selection for drought tolerance (Ribaut, González-de-León et al., 1997). Preliminary results suggest that molecular markers could be used to reduce both the cost and the time associated with recurrent selection for drought tolerance by focusing selection only on those portions of the genome known to be associated with drought tolerance.

Roots

The "hidden half" of the maize plant has received little attention in the quest for drought tolerance, though considerably more is known of the genetic control of rooting characteristics in rice (Nguyen, Chandra Babu, and Blum, 1997). Studies of water extraction patterns by maize show that little water is removed below 70 cm soil depth (Mugo et al., 1998), even though roots can be recovered in the same soil at depths up to 120 cm. Ideally, we would like to be able to manipulate roots so that their total biomass remained unchanged, but their distribution was altered to make it more uniform and to increase the effective depth of soil exploration.

The capacitance meter provides a quick and nondestructive estimate of the absorptive area of roots in maize (van Beem, Smith, and Zobel, 1998). We have conducted four cycles of recurrent full-sib selection for high and low root capacitance in a population, Pool 16 Sequía, in Mexico, and are presently evaluating the changes due to selection in a series of field and pot trials. Initial impressions suggest that selection for large capacitance has resulted in extensive development of fine roots, increased brace roots, and in later-flowering, larger plants. Selection for low capacitance appears to result in less branching of roots near the surface and in smaller plants. Selections are currently being evaluated under low N, drought, and unstressed conditions.

Genes and Transgenes?

There is a growing tendency to move beyond QTL to the identification of the genes that underlie their response, and currently available molecular tools can be used to greatly assist in this task. Within the next two to three years it is hoped that the major genes underlying ASI will be identified and cloned. Much attention is also being paid to genes that control the expression of late embryogenesis abundant (LEA) proteins. Dehydrins, which form part of the family of LEA proteins (Egerton-Warburton, Balsamo, and Close, 1997), can be induced by the application of ABA and are associated with desiccation tolerance. The transfer of a gene coding for an LEA protein, *HVA1*, from barley to rice increased its ability to withstand drought and salt stress (Xu et al., 1996). Dehydrin loci are found within the QTL interval on chromosome 1 controlling ASI in maize (Campbell and Close, 1997). An improved understanding of synteny will also markedly increase our capacity to identify and transfer drought-adaptive genes among crop species.

Genes coding for putative osmoprotectants, such as glycine betaine, proline, and mannitol are being identified and engineered into common crop species such as rice and tobacco (Nguyen, Chandra Babu, and Blum, 1997). The overproduction of proline in transgenic tobacco has increased its ability to withstand osmotic stress (Yoshida et al., 1997). As a note of caution, however, it appears that a number of these transgenic products which aid in survival under osmotic stress may well carry a cost to production in well-watered environments.

CONCLUSIONS: AN EFFECTIVE STRATEGY FOR IMPROVING DROUGHT TOLERANCE IN MAIZE

The following summarizes key points and describes the elements of what we believe is an effective strategy for improving drought tolerance in maize.

• A clear definition of the target environment is essential if the breeder is to decide on the relative utility of escape through earliness or tolerance to drought at seedling, vegetative, or reproductive growth stages.

• In choosing secondary traits to improve the efficiency of selection, those which contribute to productivity will be generally more useful than those which contribute mainly to survival, especially in the context of commercial grain production.

• Recurrent selection in elite maize germplasm has been effective at increasing tolerance to drought in tropical maize populations, resulting in increases in grain yield of around 80 to 100 $kg \cdot ha^{-1}$ per year. Improvements were not at the cost of yield in good conditions and were related to increases in partitioning to the ear, ear growth rate, spikelet fitness, and, hence, resistance to barrenness. Furthermore, improvements in drought tolerance result in improvements in tolerance to low N as well, and to some other stresses, such as shading. Conventional multilocation testing has resulted in improvements under favorable or mildly stressed conditions, but not where stress is severe.

• Key to this relatively high rate of progress has been the use of managed field-based stress focused on the flowering period, when maize is relatively susceptible. The stress must be severe enough so that genetic variation for tolerance can be easily detected. Data on various traits from strongly contrasting water regimes should be expressed in standard measure before being combined for use in selection. Care must be taken that selections for tolerance do not simply reflect escapes through early flowering.

• In addition to grain yield, it is recommended that a series of simple secondary traits be used in selection. In descending order of importance, these are barrenness (i.e., ears per plant); anthesis-silking interval; stay green; tassel size; and upright, unrolled leaves. Other traits, such as prolificacy and tolerance to low N per se, contribute little or nothing to drought tolerance. Traits indicative of plant water status are generally too time-consuming to measure directly in a nursery or show little relationship with grain yield under drought.

• Screening germplasm for improved tolerance to high plant density should result in some improvement in drought tolerance, since a short ASI has also been related to density tolerance in maize as well as drought tolerance (Edmeades et al., 2000).

• More rapid progress for drought tolerance can be made by national programs if screening is of elite, adapted germplasm rather than of unimproved or poorly adapted germplasm that may have enhanced survival capability but limited capacity to produce grain. A drought-tolerant cultivar must retain the capacity to perform competitively under unstressed conditions.

• New molecular tools, including markers and transgenes, show real promise in speeding the process of accumulating favorable drought-adaptive alleles in otherwise susceptible genotypes without significantly affecting their specific and general combining abilities. Progress in identifying and cloning drought tolerance genes can be expected in the next five years.

• The field remains the definitive proving ground for all drought-adaptive traits; wide testing of drought-tolerant selections under an array of water deficits is recommended to verify their benefits and their costs.

• Reality check: About 20 percent of the yield gap between water-stressed yields and yield potential as determined by radiation and temperature can probably be closed by breeding. Another 20 percent can be met by innovative water conservation practices. The remaining 60 percent will simply have to come from additional water supplies.

REFERENCES

Andrade, F.H., Cirilo, A.G., Uhart, S.A., and Otegui, M.E. (1996). *Ecofisiología del cultivo de maíz*. Balcarce, Argentina: La Barrosa, CERBAS, and Dekalb Press.

Bänziger, M., Betrán, F.J., and Lafitte, H.R. (1997). Efficiency of high-nitrogen selection environments for improving maize for low-nitrogen target environments. *Crop Science* 37: 1103-1109.

Bänziger, M., Edmeades, G.O., and Lafitte, H.R. (1999). Selection for drought tolerance increases maize yields over a range of N levels. *Crop Science* 39: 1035-1040.

Bänziger, M., Edmeades, G.O., and Quarrie, S. (1997). Drought stress at seedling stage—Are there genetic solutions? In Edmeades, G.O., Bänziger, M., Mickelson, H.R., and Peña-Valdivia, C.B. (Eds.), *Developing drought- and low-N-tolerant maize* (pp. 348-354). Mexico DF, Mexico: CIMMYT.

Bänziger, M. and Lafitte, H.R. (1997). Efficiency of secondary traits for improving maize for low-nitrogen target environments. *Crop Science* 37: 1110-1117.

Bassetti, P. and Westgate, M.E. (1993). Senescence and receptivity of maize silks. *Crop Science* 33: 275-278.

Beck, D., Betrán, J., Bänziger, M., Edmeades, G.O., Ribaut, J.-M., Willcox, M., Vasal, S.K., and Ortega, C.A. (1996). Progress in developing drought and low soil nitrogen tolerance in maize. In Wilkinson, D. (Ed.), *Proceedings 51st annual corn and sorghum research conference*. Chicago, December 10-11. Washington, DC: ASTA.

Blum, A. (1988). *Plant breeding for stress environments*. Boca Raton, FL: CRC Press.

Blum, A. (1997). Constitutive traits affecting plant performance under stress. In Edmeades, G.O., Bänziger, M., Mickelson, H.R., and Peña-Valdivia, C.B.

(Eds.), *Developing drought- and low-N-tolerant maize* (pp. 131-135). Mexico DF, Mexico: CIMMYT.

Blum, A. (1998). Improving wheat grain filling under stress by stem reserve mobilisation. *Euphytica* 100: 77-83.

Bolaños, J. and Edmeades, G.O. (1991). Value of selection for osmotic potential in tropical maize. *Agronomy Journal* 83: 948-956.

Bolaños, J. and Edmeades, G.O. (1993a). Eight cycles of selection for drought tolerance in lowland tropical maize. I. Responses in grain yield, biomass, and radiation utilization. *Field Crops Research* 31: 233-252.

Bolaños, J. and Edmeades, G.O. (1993b). Eight cycles of selection for drought tolerance in lowland tropical maize. II. Responses in reproductive behavior. *Field Crops Research* 31: 253-268.

Bolaños, J. and Edmeades, G.O. (1996). The importance of the anthesis-silking interval in breeding for drought tolerance in tropical maize. *Field Crops Research* 48: 65-80.

Bolaños, J., Edmeades, G.O., and Martinez, L. (1993). Eight cycles of selection for drought tolerance in lowland tropical maize. III. Responses in drought—Adaptive physiological and morphological traits. *Field Crops Research* 31: 269-286.

Boyer, J.S. (1996). Advances in drought tolerance in plants. *Advances in Agronomy* 56: 187-218.

Byrne, P.F., Bolaños, J., Edmeades, G.O., and Eaton, D.L. (1995). Gains from selection under drought versus multilocation testing in related tropical maize populations. *Crop Science* 35: 63-69.

Campbell, S.A. and Close, T.J. (1997). Dehydrins: Genes, proteins, and associations with phenotypic traits. *New Phytologist* 137: 61-74.

Castleberry, R.M., Crum, C.W., and Krull, C.F. (1984). Genetic improvements of U.S. maize cultivars under varying fertility and climatic environments. *Crop Science* 24: 33-36.

Castleberry, R.M. and Lerette, R.J. (1979). *Latente*, a new type of drought tolerance? In Loden, H. and Wilkinson, D. (Eds.), *Proceedings 34th annual corn and sorghum research conference* (pp. 46-56). Chicago, December 11-13. Washington, DC: ASTA.

Chapman, S.C., Crossa, J., and Edmeades, G.O. (1997). Genotype by environment effects and selection for drought tolerance in tropical maize. I. Two mode pattern analysis of yield. *Euphytica* 95: 1-9.

Chapman, S.C. and Edmeades, G.O. (1999). Selection improves drought tolerance in tropical maize populations. II. Direct and correlated responses among secondary traits. *Crop Science* 39: 1315-1324.

Chimenti, C.A., Cantagallo, J., and Guevara, E. (1997). Osmotic adjustment in maize: Genetic variation and association with water uptake. In Edmeades, G.O., Bänziger, M., Mickelson, H.R., and Peña-Valdivia, C.B. (Eds.), *Developing drought- and low-N-tolerant maize* (pp. 200-204). Mexico DF, Mexico: CIMMYT.

Duvick, D.N. (1997). What is yield? In Edmeades, G.O., Bänziger, M., Mickelson, H.R., and Peña-Valdivia, C.B. (Eds.), *Developing drought- and low-N-tolerant maize* (pp. 332-335). Mexico DF, Mexico: CIMMYT.

Edmeades, G.O., Bänziger, M., Chapman, S.C., Ribaut, J.-M., and Bolaños, J. (1997). Recent advances in breeding for drought tolerance in maize. In Badu-Apaku, B., Akoroda, M.O., Ouedraogo, M., and Quin, F.M. (Eds.), *Contributing to food self-sufficiency: Maize research and development in West and Central Africa* (pp. 22-41). Proceedings of the West and Central Africa Regional Maize Workshop, Cotonou, Benin, May 28-June 2, 1995. Ibadan, Nigeria: IITA.

Edmeades, G.O., Bänziger, M., Chapman, S.C., Ribaut, J.-M., and Elings, A. (1997). Recent advances in breeding for drought tolerance in maize. In Kropff, M.J., Teng, P.S., Aggarwal, P.K., Bouma, J., Bouman, B.A.M., Jones, J.W., and van Laar, H.H. (Eds.), *Applications of systems approaches at the field level* (pp. 63-78). Proceedings of the Second International Symposium on Systems Approaches for Agricultural Development (SAAD2). IRRI, Los Baños, Philippines, December 6-8, 1995. Dordrecht, Netherlands: Kluwer Academic Publishers.

Edmeades, G.O., Bänziger, M., Cortes, C.M., and Ortega, C.A. (1997). From stress-tolerant populations to hybrids: The role of source germplasm. In Edmeades, G.O., Bänziger, M., Mickelson, H.R., and Peña-Valdivia, C.B. (Eds.), *Developing drought- and low-N-tolerant maize* (pp. 263-273). Mexico DF, Mexico: CIMMYT.

Edmeades, G.O., Bolaños, J., Bänziger, M., Chapman, S.C., Ortega, C.A., Lafitte, H.R., Fischer, K.S., and Pandey, S. (1997). Recurrent selection under managed drought stress improves grain yields in tropical maize. In Edmeades, G.O., Bänziger, M., Mickelson, H.R., and Peña-Valdivia, C.B. (Eds.), *Developing drought- and low-N-tolerant maize* (pp. 415-425). Mexico DF, Mexico: CIMMYT.

Edmeades, G.O., Bolaños, J., Bänziger, M., Ribaut, J.-M., White, J.W., Reynolds, M.P., and Lafitte, H.R. (1998). Improving crop yields under water deficits in the tropics. In Chopra, V.L., Singh, R.B., and Varma, A. (Eds.), *Crop productivity and sustainability—Shaping the future* (pp. 437-451). Proceedings of the Second International Crop Science Congress. New Delhi, India: Oxford and IBH.

Edmeades, G.O., Bolaños, J., and Chapman, S.C. (1997). Value of secondary traits in selecting for drought tolerance in tropical maize. In Edmeades, G.O., Bänziger, M., Mickelson, H.R., and Peña-Valdivia, C.B. (Eds.), *Developing drought- and low-N-tolerant maize* (pp. 222-234). Mexico DF, Mexico: CIMMYT.

Edmeades, G.O., Bolaños, J., Chapman, S.C., Lafitte, H.R., and Bänziger, M. (1999). Selection improves drought tolerance in tropical maize populations. I. Gains in biomass, grain yield and harvest index. *Crop Science* 39: 1306-1315.

Edmeades, G.O., Bolaños, J., Elings, A., Ribaut, J.-M., Bänziger, M., and Westgate, M.E. (2000). The role and regulation of the anthesis-silking interval in maize. In Mickelson, S. (Ed.), *Physiology and modeling kernel set in maize,* Special publication number 29 (pp. 43-73). Madison, WI: Crop Science Society of America.

Edmeades, G.O., Bolaños, J., Hernandez, M., and Bello, S. (1993). Causes for silk delay in lowland tropical maize. *Crop Science* 33: 1029-1035.

Edmeades, G.O., Bolaños, J., and Lafitte, H.R. (1992). Progress in breeding for drought tolerance in maize. In Wilkinson, D. (Ed.), *Proceedings 47th annual corn and sorghum industry research conference*, (pp. 93-111). Chicago, December 8-10, 1992. Washington, DC: ASTA.

Edmeades, G.O. and Daynard, T.B. (1979). The relationship between final yield and photosynthesis at flowering in individual maize plants. *Canadian Journal of Plant Science* 59: 585-601.

Edmeades, G.O. and Lafitte, H.R. (1993). The effects of defoliation and high plant density stress on tropical maize selected for reduced plant height. *Agronomy Journal* 85: 850-857.

Egerton-Warburton, L.M., Balsamo, R.A., and Close, T.J. (1997). Temporal accumulation and ultrastructural location of dehydrins in *Zea mays*. *Physiologia Plantarum* 101: 454-555.

Eyhérabide, G.H., Guevara, E., and Totis de Zeljkovich, L. (1997). Efecto de estrés hídrico en al rendimiento de maíz en Argentina. In Edmeades, G.O., Bänziger, M., Mickelson, H.R., and Peña-Valdivia, C.B. (Eds.), *Developing drought- and low-N-tolerant maize* (pp. 24-28). Mexico DF, Mexico: CIMMYT.

Fischer, K.S., Edmeades, G.O., and Johnson, E.C. (1987). Recurrent selection for reduced tassel branch number and reduced leaf area density above the ear in tropical maize populations. *Crop Science* 27: 1150-1156.

Fischer, K.S., Johnson, E.C., and Edmeades, G.O. (1983). *Breeding and selection for drought resistance in tropical maize*. El Batan, Mexico: CIMMYT.

Fischer, R.A. (1989). Cropping systems for greater drought resistance. In Baker, F.W.G. (Ed.), *Drought resistance in cereals* (pp. 201-211). Paris and Wallingford: ICSU and CABI.

Fukai, S. and Cooper, M. (1995). Development of drought-resistant cultivars using physio-morphological traits in rice. *Field Crops Research* 40: 67-86.

Grant, R.F., Jackson, B.S., Kiniry, J.R., and Arkin, G.F. (1989). Water deficit timing effects on yield components in maize. *Agronomy Journal* 81: 61-65.

Hall, A.E., Villela, F., Trapani, N., and Chimenti, C. (1982). The effects of water stress and genotype on the dynamics of pollen shedding and silking in maize. *Field Crops Research* 5: 349-363.

Hauck, B., Gay, A.P., MacDuff, J., Griffiths, C.M., and Thomas, H. (1997). Leaf senescence in a non-yellowing mutant of *Festuca pratensis:* Implications of the *stay-green* mutation for photosynthesis, growth and nitrogen nutrition. *Plant, Cell and Environment* 20: 1007-1018.

Heisey, P.W. and Edmeades, G.O. (1999). Maize production in moisture-stressed environments. In *World maize facts and trends, 1997-1998* (pp. 1-36). El Batan, Mexico: CIMMYT.

Jensen, S.D. and Cavalieri, A.J. (1983). Drought tolerance in U.S. maize. *Agricultural Water Management* 7: 223-236.

Lafitte, H.R and Edmeades, G.O. (1995). Stress tolerance in tropical maize is linked to constitutive changes in ear growth characteristics. *Crop Science* 35: 820-826.

Lemcoff, J.H., Chimenti, C.A., and Davezac, T.A.E. (1998). Osmotic adjustment in maize (*Zea mays* L.): Changes with ontogeny and its relationship with phenotypic stability. *Journal of Agronomy and Crop Science* 180: 241-247.

Ludlow, M.M. and Muchow, R.C. (1990). A critical evaluation of traits for improving crop yields in water-limited environments. *Advances in Agronomy* 43: 107-153.

Mugo, S.N., Smith, M.E., Bänziger, M., Setter, T.L., Edmeades, G.O., and Elings, A. (1998). Performance of early maturing Katumani and Kito maize composites under drought stress. *African Journal of Crop Science* 6: 329-344.

Nguyen, H.T., Chandra Babu, R., and Blum, A. (1997). Breeding for drought resistance in rice: Physiology and molecular genetics considerations. *Crop Science* 37: 1426-1434.

Passioura, J.B. (1977). Grain yield, harvest index, and water use of wheat. *Journal of the Australian Institute of Agricultural Science* 43: 117-120.

Reeder, L. (1997). Breeding for yield stability in a commercial program in the USA. In Edmeades, G.O., Bänziger, M., Mickelson, H.R., and Peña-Valdivia, C.B. (Eds.), *Developing drought- and low-N-tolerant maize* (pp. 387-391). Mexico DF, Mexico: CIMMYT.

Ribaut, J.-M., González-de-León, D., Jiang, C., Edmeades, G.O., and Hoisington, D. (1997). Identification and transfer of ASI quantitative trait loci (QTL): A strategy to improve drought tolerance in maize lines and population. In Edmeades, G.O., Bänziger, M., Mickelson, H.R., and Peña-Valdivia, C.B. (Eds.), *Developing drought- and low-N-tolerant maize* (pp. 396-400). Mexico DF, Mexico: CIMMYT.

Ribaut, J.-M., Hoisington, D.A., Deutsch, J.A., Jiang, C., and González-de-León, D. (1996). Identification of quantitative trait loci under drought conditions in tropical maize. I. Flowering parameters and the anthesis-silking interval. *Theoretical and Applied Genetics* 92: 905-914.

Ribaut, J.-M., Jiang, C., González-de-León, D., Edmeades, G.O., and Hoisington, D.A. (1997). Identification of quantitative trait loci under drought conditions in tropical maize. II. Yield components and marker-assisted selection strategies. *Theoretical and Applied Genetics* 94: 887-896.

Richards, R.A. (1991). Crop improvement for temperate Australia: Future opportunities. *Field Crops Research* 26: 141-169.

Richards, R.A. (1992). The effect of dwarfing genes in spring wheat in dry environments. II. Growth, water use and water use efficiency. *Australian Journal of Agricultural Research* 43: 529-539.

Sadras, V.O. and Connor, D.J. (1991). Physiological basis of the response of harvest index to the fraction of water transpired after anthesis: A simple model to estimate harvest index for determinate species. *Field Crops Research* 26: 227-239.

Saini, H.S. and Lalonde, S. (1998). Injuries to reproductive development under water stress, and their consequences for crop productivity. *Journal of Crop Production* 1: 223-248.

Schoper, J.B., Lambert, R.J., and Vasilas, B.L. (1986). Maize pollen viability and ear receptivity under water and high temperature stress. *Crop Science* 26: 1029-1033.

Schussler, J.R. and Westgate, M.E. (1995). Assimilate flux determines kernel set at low water potential in maize. *Crop Science* 35: 1074-1080.

Setter, T.L. (1997). Roles of the phytohormone ABA in drought tolerance: Potential utility as a selection tool. In Edmeades, G.O., Bänziger, M., Mickelson, H.R., and Peña-Valdivia, C.B. (Eds.), *Developing drought- and low-N-tolerant maize* (pp. 142-150). Mexico DF, Mexico: CIMMYT.

Shaw, R.H. (1977). Water use and requirements of maize—A review. In *Agrometeorology of the maize (corn) crop* (pp. 119-134). Geneva, Switzerland: Secretariat, World Meteorological Organization, No. 481.

Sinclair, T.R., Bingham, G.E., Lemon, E.R., and Allen, L.H. (1975). Water use efficiency of field-grown maize during moisture stress. *Plant Physiology* 56: 245-249.

Smart, C.M., Hosken, S.E., Thomas, H., Greaves, J., Blair, B.G., and Schuch, W. (1995). The timing of maize leaf senescence and characterisation of senescence-related cDNAs. *Physiologia Plantarum* 93: 673-682.

Subbarao, G.V., Johansen, C., Slinkard, A.E., Nageswara Rao, R.C., Saxena, N.P., and Chauhan, Y.S. (1995). Strategies for improving drought resistance in grain legumes. *Critical Reviews in Plant Sciences* 14: 469-523.

Turner, N.C., O'Toole, J.C., Cruz, R.T., Namuco, O.S., and Ahmed, S. (1986). Responses of seven diverse rice cultivars to water deficits. I. Stress development, canopy temperature, leaf rolling and growth. *Field Crops Research* 13: 257-271.

van Beem, J., Smith, M.E., and Zobel, R.W. (1998). Estimating root mass in maize using a portable capacitance meter. *Agronomy Journal* 90: 566-570.

Walulu, R.S., Rosenow, D.T., Webster, D.B., and Nguyen, H.T. (1994). Inheritance of stay-green trait in sorghum. *Crop Science* 34: 970-972.

Westgate, M.E. (1997). Physiology of flowering in maize: Identifying avenues to improve kernel set during drought. In Edmeades, G.O., Bänziger, M., Mickelson, H.R., and Peña-Valdivia, C.B. (Eds.), *Developing drought- and low-N-tolerant maize* (pp. 136-141). Mexico DF, Mexico: CIMMYT.

Westgate, M.E. and Boyer, J.S. (1986). Reproduction at low silk and pollen water potentials in maize. *Crop Science* 26: 951-956.

Xu, D., Duan, X., Wang, B., Hong, B., Ho, T.D., and Wu, R. (1996). Expression of a late embryogenesis abundant protein gene, *HVA1*, from barley confers tolerance to water deficit and salt stress in transgenic rice. *Plant Physiology* 110: 249-257.

Yoshida, Y., Kiyosue, T., Nakashima, K., Yamaguchi-Shinozaki, K., and Shinozaki, K. (1997). Regulation of levels of proline as an osmolyte in plants under water stress. *Plant and Cell Physiology* 38: 1095-1102.

Zinselmeier, C., Schussler, J.R., Westgate, M.E., and Jones, R.J. (1995). Low water potential disrupts carbohydrate metabolism in maize ovaries. *Plant Physiology* 107: 385-391.

Chapter 7

Strategies to Maintain Ovary and Kernel Growth During Drought

Mark E. Westgate

INTRODUCTION

In maize (*Zea mays* L.), most yield losses due to drought during reproductive development result from a decrease in the number of ears per plant, a decrease in kernels per ear, or production of smaller kernels, depending on when the drought occurs. Loss of pollen viability, silk receptivity, and lack of fertilization after pollination also can contribute to the yield loss. Failure to maintain rapid ovary and kernel growth, however, is the primary cause for reproductive failure. This chapter examines factors known to affect the rate and duration of ovary and kernel growth as a basis for developing strategies to improve yield of maize during drought.

Failure to produce kernels during drought stems from maize having separate staminate and pistillate flowers. Drought inhibits ear and silk growth, causing a delay in silk emergence relative to pollen shed (protandry) (DuPlessis and Dijkhuis, 1967; Edmeades et al., 1993). The typical breeding strategy to minimize this problem has been to select against protandry and for high yield across environments. Selection for a minimum anthesis-silking interval (ASI) under controlled drought conditions has also proven successful (Edmeades et al., 1997). Recent evidence indicates this approach is an indirect selection for rapid ear growth prior to and during anthesis. If pollination does occur during drought, the newly formed zygotes may abort within a few days of pollination (Westgate and Boyer, 1986a). The number of kernels that develop on the ear of maize is closely correlated with the rate of ovary growth during pollination. Ear and early kernel development are highly dependent on the flux of assimilates from the leaves, which is severely limited during drought due to the inhibition of photosynthesis and poor "sink strength" of the ear. Thus, increasing the partitioning of available sucrose to the ear is fundamental in strategies designed to minimize kernel abortion during drought.

Drought during grain filling causes maize to produce smaller kernels, which results primarily from the premature cessation of kernel growth (Westgate, 1994). Several studies suggest that desiccation, a normal process in kernel development, ultimately causes kernel growth to cease in both well-watered and droughted plants. If so, delaying the onset of kernel desiccation is critical for maintaining kernel growth during drought.

EAR AND OVARY GROWTH PRIOR TO POLLINATION

Close synchrony between silk emergence and pollen shed is essential for maximizing kernel set during drought. Edmeades and colleagues (1997) have shown that performance under severe drought conditions can be improved by selection in a droughty environment. The yield gain is correlated with an increase in kernel-bearing ears and a shorter anthesis-silking interval (ASI) (see Table 7.1).

Using a quantitative approach, Bassetti and Westgate (1994) determined that kernel loss due to an increase in ASI was primarily a consequence of failure of silks to emerge rather than loss of silk receptivity or lack of sufficient pollen. Distinguishing between these three possibilities is important because the selection strategies for improving kernel set under drought conditions would be markedly different in each case. Accelerating silk emergence relative to pollen shed, for example, focuses on increasing assimilate partitioning to the ear prior to anthesis.

Under well-watered conditions, the ASI is generally about one day (see Figure 7.1). Note that the most intense period of pollen shed occurs about three days after anthesis. It is not surprising, then, that maximum kernel set was obtained on ears whose silks first appeared within two or three days of anthesis (see Figure 7.2). Percent set decreased on ears with later-emerging silks, as kernels were lost

TABLE 7.1. Yield and Associated Traits for Cycle 0 and Cycle 8 of Tuxpeño Sequía Selected for a Short ASI Under Severe Drought Conditions

Tuxpeño Sequía	Yield droughted (t·ha⁻¹)	Yield well-watered (t·ha⁻¹)	ASI droughted (days)	Prolificacy droughted (ears/ plant)	Harvest index droughted (t·t⁻¹)
C0	1.75	7.48	6.4	0.73	0.12
C8	2.39	7.78	2.9	0.93	0.22

Source: Adapted from Edmeades et al. (1997, p. 419).

progressively from tip to basal florets with increasing ASI (data not shown). Related measurements of kernel distribution on open-pollinated ears indicated that sufficient pollen was available for nearly perfect kernel set of florets whose silks had emerged up to seven days after anthesis (Bassetti and Westgate, 1994). With ASI greater than seven days, both silk emergence and pollen shed limited kernel set. A negative ASI (silk emergence prior to pollen shed) as long as six days caused no loss in kernel production under favorable conditions (see Figure 7.2). Such a large negative ASI could be advantageous under drought conditions because any delay in silk emergence would actually improve the synchrony between maximum pollen shed and silk emergence. This conclusion is supported by Moser and co-workers (1997) who associated increased kernel number per ear during drought in La Posta Sequía C₄ with a negative ASI of two and a half days under well-watered conditions.

A short (or negative) ASI ultimately is coupled to rapid ear growth. The dramatic increase in kernel production in Tuxpeño Sequía C8 (see Table 7.1), a population selected for a short ASI under droughty conditions, is associated with an increase in spikelet biomass at anthesis (see Figure 7.3). Tuxpeño Sequía C8, selected through eight cycles for short ASI during drought (see Table 7.1), exhibits a greater rate of biomass accumulation per spikelet at anthesis than its C0 counterpart (Edmeades et al., 1993). Thus, Edmeades and his co-workers effectively selected for increased spikelet sink strength, which maintained sucrose flux to the ear and spikelet development during drought. Frederick and colleagues (1989) also associ-

FIGURE 7.1. Close Synchrony in Flower Development in Maize

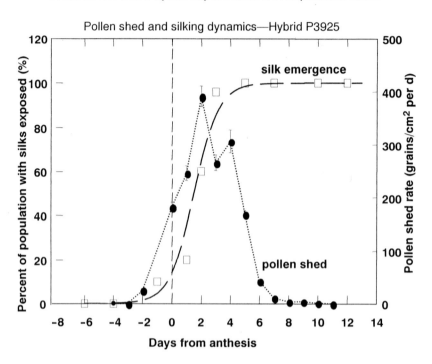

Source: M. Westgate (unpublished data).

Note: Percent of population with silks exerted is plotted relative to anthesis.

ated improved tolerance to water deficits among maize hybrids with greater ear biomass at anthesis.

OVARY GROWTH DURING POLLINATION

Under severe drought, newly formed kernels often abort within a few days of pollination (Westgate and Boyer, 1986a). Several lines of evidence indicate kernel abortion results from a lack of carbohydrate supply to the developing ear, coupled with an inhibition of carbohydrate metabolism at low ovary water potential (Boyle, Boyer, and Morgan, 1991; Schussler and Westgate, 1991; Zinselmeier et al.,

FIGURE 7.2. Percent Kernel Set on Ears with Silks Emerged at Various Times Relative to Anthesis

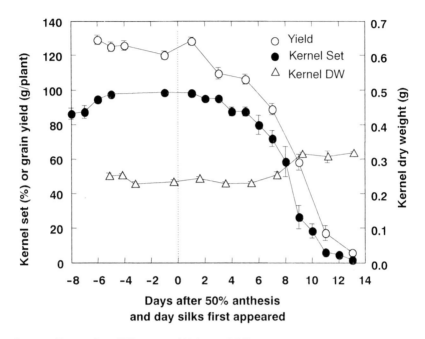

Source: Bassetti and Westgate (1994, p. 702).

TABLE 7.2. Effect of Water Deficit and Light Reduction Treatments at Pollination on Silk Water Potential and Kernel Set

Treatment	n	Silk water potential (MPa)	Kernel number per ear
Control	5	−0.41 ± 0.03	598 ± 17
−Light	11	−0.40 ± 0.04	222 ± 72
−Water	8	−1.00 ± 0.02	15 ± 6

Source: Schussler and Westgate (1991b, p. 1198).

FIGURE 7.3. Relationship Between ASI and Spikelet Biomass at Anthesis for Tuxpeño Sequía

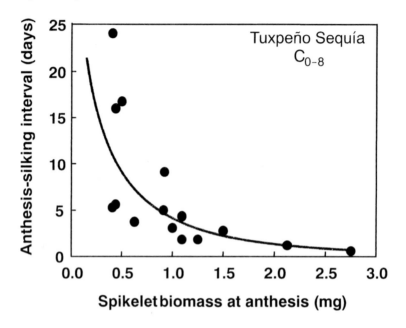

Source: Edmeades et al. (1993, p. 1033).

1995). Shading well-watered plants to inhibit leaf photosynthetic rates to the same extent as occurs in droughted plants accounts for about 70 percent of the kernel loss caused by drought (see Table 7.2).

Similarly, infusing a solution containing sucrose into stems of growth chamber plants droughted during pollination can recover about 70 percent of the kernels lost during drought (Zinselmeier et al., 1995). Together, these results confirm that early reproductive development in maize is closely coupled to the supply of photosynthate delivered to the ear. Unlike the situation during kernel filling, however, lack of current photosynthate at anthesis is not buffered by mobilization of reserve carbohydrates accumulated earlier in development. It has not been possible to decrease the sensitivity of kernel set to drought by increasing the level of reserves culturally

(Schussler and Westgate, 1994) or by increasing source:sink ratios genetically (Zinselmeier, Westgate, and Jones, 1995) prior to pollination. This situation is very problematic because the photosynthetic machinery remains very sensitive to low leaf water potentials, even in modern hybrids tolerant to high population densities. A possible alternative for maintaining flux of carbohydrate to the ear when photosynthesis is inhibited is to modify the process of sugar storage in the stem. The stem, shank, and cob of maize accumulate large amounts of sugar during elongation and maturation (see Figure 7.4.), which coincide with the period of rapid ear and ovary growth (Wendler et al., 1991; Schussler and Westgate, 1991a; Schussler and Westgate, 1995). Most of the sugar that accumulates in the elongating stem internodes, shank, and cob is in the form of glucose, fructose, and/or starch, which are not remobilized when photosynthesis is inhibited. Relative to the ear, these structures are strong sinks for carbohydrates that can continue to accumulate sugars at the expense of the ear when drought occurs at anthesis (Schussler and Westgate, 1991b; Schussler and Westgate, 1994). Sucrose accumulates after anthesis, when elongation is complete (Reed and Singletary, 1989; Zinselmeier, 1991). Nearly all this sucrose can be remobilized to support grain development if photosynthesis is inhibited during grain filling (Jones and Simmons, 1983; Westgate, 1994). Our current studies are aimed at increasing the proportion of sucrose in the stem, shank, and cob tissues at anthesis in an attempt to maintain the flux of sucrose to the ear when photosynthesis is inhibited by drought.

Not all the kernel abortion observed on droughted plants in the studies of Schussler and Westgate (1991b); Boyle, Boyer, and Morgan (1991); and Zinselmeier and colleagues (1995) could be accounted for by lack of concurrent photosynthate supply to the ear. Zinselmeier, Lauer, and Boyer (1995) discovered that the activities of the soluble and insoluble acid invertase were inhibited in ovaries of plants that were droughted severely enough to cause ovary abortion (see Figure 7.5). Related measurements showed that the ratio of glucose + fructose to sucrose in the ovary free space decreased dramatically in droughted ovaries, consistent with an inhibition of insoluble invertase activity in the pedicel (M. Westgate, unpublished data). Evidently, drought caused a lesion in carbohydrate metabolism within the developing ovaries that diminished their capacity to utilize

Figure 7.4. Reducing Sugar and Sucrose Levels in Cob, Shank, and Internode Tissues of Maize Hybrid A629 × W64a During Pollination

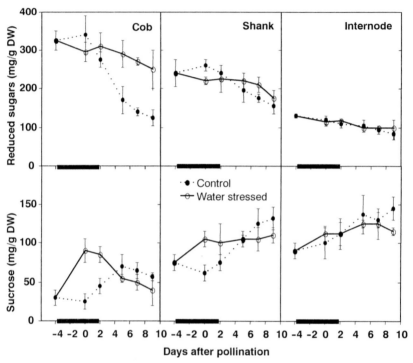

Source: C. Zinselmeier (unpublished data).

Note: Water was withheld at silk emergence (horizontal bars) for six days.

sucrose supplied by the plant. A similar phenomenon has been reported during abortion of pollen grains in droughted wheat (Saini, 1997). The molecular basis for the inhibition of invertase activity is not known but could be caused by sugar modulation of invertase genes (Koch et al., 1992). The most important consequence of this metabolic lesion is an inhibition of ovary growth (see Figure 7.6).

In a wide range of studies examining why drought increases kernel abortion in maize, final kernel numbers per ear were directly correlated with the inhibition of ovary growth regardless of genotype, potential

FIGURE 7.5. Specific Activity of Cell Wall-Bound Acid Invertase from Maize Ovaries at Decreasing Ovary Water Potential

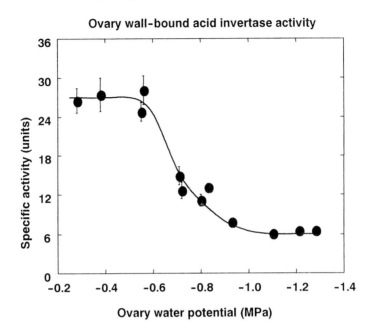

Source: M. Westgate (unpublished data).

Note: Kernels fail to develop from ovaries pollinated at −1.0 MPa or lower.

kernel number per ear, culture conditions, or plant pretreatment (Schussler and Westgate, 1995). Whether accomplished artificially (Boyle, Boyer, and Morgan, 1991; Zinselmeier, Lauer, and Boyer, 1995), culturally (Schussler and Westgate, 1994; Zinselmeier, Westgate, and Jones, 1995), or genetically (Edmeades et al., 1993; Edmeades et al., 1997), improvements in kernel set under drought conditions have been coupled closely with the maintenance of ovary growth. The association between improvements in performance of maize in droughty environments and increased rate of ovary growth during pollination (Bolaños and Edmeades, 1993; Edmeades et al., 1993) provides direct support for this conclusion.

FIGURE 7.6. Correlation Between Kernel Number per Ear and Relative Ovary Growth Rate During Pollination

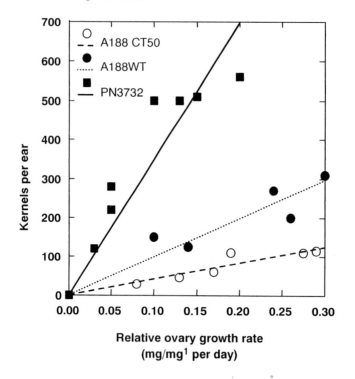

Source: Adapted from Schussler and Westgate (1991a), Schussler and Westgate (1995), and Zinselmeier et al. (1995).

Note: P3732, maize hybrid; A188WT, maize inbred; A188TM, translocation mutant of A188WT with 50 percent ovary abortion. Data are pooled from shading and drought treatments using growth chamber and field-grown plants.

KERNEL GROWTH AFTER POLLINATION

The impact of drought on kernel growth depends on the stage of kernel development at which the drought occurs. As such, the three stages of kernel growth described by Bewley and Black (1994) provide a useful framework for this discussion. Stage I, often termed the "lag phase," is a period of cell division and differentiation marked by a rapid increase in kernel fresh weight. Enlargement is primarily the result of water influx driven by a rapid accumulation of solutes (Barlow et al., 1980; Westgate and Boyer, 1986b). Under favorable growing conditions, potential kernel mass is determined during this period by the number of endosperm cells that form (Jenner, Ugalde, and Aspinall, 1991; Jones, Schreiber, and Roessler, 1996). Overlapping and following is Stage II, grain filling. This period is marked by a rapid gain in kernel dry weight as a result of the synthesis and deposition of the stored reserves. Starch is the predominant storage material in the endosperm and constitutes the bulk of the mature kernel mass. Kernel fresh weight remains relatively stable, although the total water content declines as water is displaced by the reserves accumulating within the cells of the embryo and endosperm. Dry-matter accumulation ceases during Stage III as the kernel undergoes maturation drying and approaches a "quiescent state." Loss in fresh weight reflects a continued, and sometimes more rapid, decline in water content. Accumulation of late embryogenesis abundant (LEAs) proteins and carbohydrates during this stage is thought to protect cellular membranes from the desiccation that occurs in mature kernels (Dure, 1997).

Numerous studies have shown that drought during these three stages of kernel development decreases seed set and/or kernel size. As detailed earlier, water deficits during pollination cause cessation of embryo and endosperm abortion within a few days of fertilization (Westgate and Boyer, 1986a). Lack of water from five to eleven days after pollination (Stage I) also exerts its effect primarily on kernel set (Artlip, Madison, and Setter, 1995; Mambelli and Setter, 1998). In this case, kernels grow sufficiently to allow some accumulation of starch and zein in the embryo and endosperm, but development is aborted prematurely, and kernels are incompletely filled (NeSmith and Ritchie, 1992). Water deficits during Stage II generally have little impact on kernel set but can decrease final kernel size (Ouattar, Jones, and Crookston, 1987; Grant et al., 1989; Westgate, 1994). In most cases, drought shortens the duration of the filling period, but the rate of filling also can be affected under severe drought conditions (Jurgens, Johnson, and Boyer, 1978; Westgate, 1994).

Stage I: Reversing the Loss of Sink Capacity

Sink capacity is a function of both the number of endosperm cells and the number of starch grains initiated within the amyloplasts of those cells (Jones, 1994). Water deficits as well as high temperatures during this stage reduce endosperm cell number (Jones, Roessler, and Ouattar, 1985; Ouattar, Jones, and Crookston, 1987; Artlip, Madison, and Setter, 1995) and, therefore, decrease the number of sites for starch synthesis. Although the physiological mechanism that establishes the number of sites for starch synthesis is not known, it likely involves factors that control cell division and proliferation of nuclear or plastic DNA (deoxyribonucleic acid). Circumstantial evidence indicates that increased cytokinin content can enhance grain set and sometimes increase yield in maize (Dietrich et al., 1995). Also, Cheikh and Jones (1994) reported that stem infusion of benzylaminopurine (BAP) partially relieved kernel abortion caused by heat stress during Stage I. Drought imposed early in this period inhibits endoreduplication in the maize endosperm (Artlip, Madison, and Setter, 1995). Imposed later in Stage I, drought has less impact on DNA proliferation but continues to inhibit cell division and to limit kernel sink potential. Circumstantial evidence from endosperm defective (*dek*) mutants indicates that endosperm auxin content has a direct effect on the extent of endoreduplication (Lur and Setter, 1993). Currently, however, no direct evidence shows that cytokinin or auxin levels in kernels are altered by drought, or that maintaining the level of these phytohormones would be a fruitful approach to maintaining kernel growth. Further research is needed to determine why cell division and DNA synthesis are so vulnerable to drought during this stage of kernel development.

Stage II: Maintaining the Rate of Reserve Deposition

Once sink potential has been established and the kernel begins to accumulate starch and protein reserves (Stage II), drought can decrease final kernel size by limiting the rate and duration of reserve deposition. Typically, water deficits have little impact on the rate of kernel growth rate (Ouattar, Jones, and Crookston, 1987; Westgate, 1994). High temperatures that accompany drought, however, often increase kernel growth rate (Egli, 1994). Both components of drought cause premature cessation of kernel filling. Lack of assimilate supply due to leaf senescence (Jurgens, Johnson, and Boyer, 1978), decreased capacity for assimilation within the endosperm (Keeling et al., 1994), and premature desiccation of the kernel (Westgate, 1994) all can contribute to the production of smaller kernels under drought conditions.

Assimilate reserves are mobilized from the stalk and rachis to support kernel growth during drought (Jurgens, Johnson, and Boyer, 1978; Westgate and Boyer, 1985; Schussler and Westgate, 1994), and water deficits reportedly have little impact on photosynthate transport in the phloem from the peduncle (or stalk) to the kernels (Fisher and Gifford, 1986). Furthermore, estimates of sucrose concentration in the apoplast of the maize pedicel (Porter, Kneivel, and Shannon, 1985) are nearly tenfold greater than those required for maximum growth of kernels in vitro (Cobb et al., 1988). Thus, several mechanisms contribute to the stability of kernel growth rate during short-term changes in assimilate supply. Together, they function to ensure that the supply of assimilates from the plant probably does not limit the rate of reserve deposition in the kernel.

The observed stability of kernel growth across environments implies that the rate of kernel growth is determined genetically and is moderated by conditions within the kernel itself. Maize kernels exhibit little, if any, change in water potential when drought occurs during rapid grain filling, while other plant structures undergo large decreases in tissue water potential (Ouattar et al., 1987; Westgate and Grant, 1989; Westgate, 1994). Various anatomical, physico-chemical, and theoretical models have been proposed to explain the apparent hydraulic isolation of the kernels from other structures of the plant. These include vascular discontinuities within the grain (Brooks, Jenner, and Aspinall, 1982), favorable water status within the stem (Ouattar et al., 1987), osmotic regulation in the apoplast (Westgate and Boyer, 1986b), and specialized tissues within the vasculature that control osmotic potential of the apoplast (Bradford, 1994). Presumably, maintenance of a favorable water status within the kernel permits metabolism to continue despite severe water deficits in the vegetative tissues.

Stage III: Extending the Duration of Grain Filling

A shorter duration of filling is the predominant reason maize produces smaller kernels when drought occurs during the filling period (Frederick et al., 1989; Westgate, 1994). Why kernels cease to accumulate dry matter is not fully understood. Several authors have concluded that kernel growth in droughted plants continues until available assimilates, from concurrent photosynthesis, remobilization of reserves, or dismantling of assimilatory proteins, are depleted (Passioura, 1994; TeKrony and Hunter, 1995). Water deficits and high temperatures often hasten leaf senescence, thereby limiting assimilate supply late in kernel development (Jurgens, Johnson, and Boyer, 1978; Westgate and Boyer, 1985; Cirilo and Andrade, 1996). Maize hybrids with a greater capacity for assimilation late in development (i.e., with the "stay-green" character) may be more tolerant to drought during grain filling (Frederick et al., 1989). These studies, however, often imposed treatments that rapidly and severely inhibited leaf photosynthesis. Such a rapid change in assimilate supply would be highly unlikely in the field. Typically, the rate of photosynthesis decreases gradually as drought develops, and grain growth is supported by a combination of concurrent and stored assimilates from the plant. Even under these conditions, drought still shortens the duration of kernel growth. Reserve deposition, however, ceases well before reserves are completely depleted (Westgate, 1994).

Egli (1998) proposed that the cessation of cell expansion ultimately determines the subsequent pattern of seed desiccation and maturation. The maximum water volume, established at the end of Stage I, could determine final kernel size in two ways. First it sets an upper limit on cellular volume available for dry-matter accumulation. Second, it starts the "developmental clock" for grain fill duration, since percent kernel moisture decreases immediately and steadily thereafter. Grain filling in maize and other cereals apparently continues until metabolism is limited by low moisture content.

Data from a number of unrelated studies on kernel development in cereals show a general correspondence between final kernel dry weight and maximum water content during grain filling (see Figure 7.7). The correlation between kernel dry weight and maximum water content for wheat (see Figure 7.7A) is fairly strong ($r^2 = 0.86$).

FIGURE 7.7: Relationships Between Final Kernel Weight and Maximum Water Content in Several Cereals

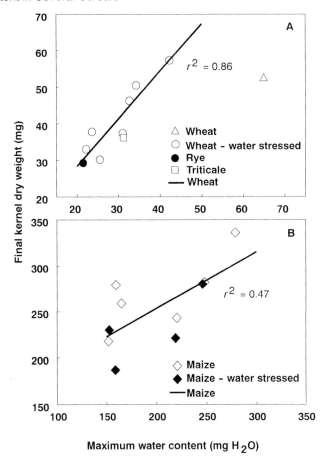

Source: Adapted from Saini and Westgate (2000).

Note: A: maize; B: wheat, rye, and triticale. Open symbols indicate kernels water stressed during grain filling.

Relationships for maize, rye, and triticale, however, are more tenuous. Triticale kernels, for example, accumulate much less dry matter than expected from their maximum water volume, and maize kernels achieved a wide range of dry weights from a maximum water content of about 150 milligrams (mg) (see Figure 7.7B). It is important to note that kernels from water-stressed plants followed the same general pattern as those from well-watered plants.

As reserves are deposited within the kernel, starch, protein, and lipid replace water volume. Although the accumulation of these reserves has no direct impact on metabolism, it effectively desiccates the endosperm and embryo. It has been suggested that desiccation late in kernel development ultimately limits the synthesis of storage reserves (Adams and Rime, 1980; Egli, 1998). If so, the general pattern of kernel dry-matter accumulation across species, genotypes, and environments should be fairly similar when expressed on a kernel moisture basis. Figure 7.8 shows the pattern of dry-weight accumulation in maize and wheat kernels with decreasing kernel moisture content.

Data were taken from four independent and unrelated studies in which kernel water content was monitored during development. The large variation in final dry weight reflects differences in potential kernel size among the species as well as different growing conditions for each experiment. Viewed in this way, it becomes quite evident that kernel dry-matter accumulation continues rapidly until kernels reach about 40 percent moisture (dry weight [DW] basis). The consistency across species and environments suggests there may indeed be a minimum moisture content beyond which reserve accumulation cannot continue.

Kernel development data from a number of studies encompassing a range of treatment conditions and species support this conclusion (see Table 7.3). The moisture content at which reserve deposition ceased (maximum [Max] DW) in these studies was estimated from simple regression analysis of DW versus kernel moisture. Plants were grown in the field, in the greenhouse, under well-watered conditions, or exposed to drought. Despite the large variation in final kernel DW among the species and treatments, and the relatively crude method used to estimate kernel moisture at Max DW, the variation in kernel moisture at Max DW is remarkably small (mean ± SD, 39 ± 2.5 percent). Of importance, this value is essentially the same for kernels developing on droughted plants. Dry-matter accumulation ceased at about 40 percent moisture in Pioneer Brand 3987 grown under well-watered conditions, or when exposed to drought as

FIGURE 7.8. Dry-Matter Accumulation in Kernels of Maize and Wheat As Percent Moisture Declines

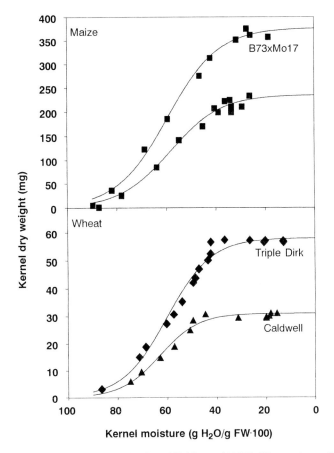

Source: Data are adapted from Egli and TeKrony (1997), Westgate and Boyer (1986b), Sofield et al. (1977), and Brooks, Jenner, and Aspinall (1982).

early in grain filling as the blister stage, or as late as first dent. Such results indicate drought does not alter the relationship between dry-matter accumulation and kernel moisture. Our field results indicate kernels on droughted plants cease to accumulate dry matter earlier

TABLE 7.3. Estimate of Kernel Moisture at Maximum Kernel Dry Weight (Max DW) for Four Cereal Species

Species	Maximum kernel DW (mg)	Regression analysis DW versus percent moisture			Kernel moisture at Max DW (percent)	Source
		B0	B1	r		
Maize						
B73xMo17	363 ± 10.0	-6.3	567	-0.99	33	Westgate and Boyer (1986b)
B73xMo17	214 ± 12.3	-4.6	388	-0.99	38	Egli and TeKrony (1997)
P3987	279 ± 15.7	-4.8	455	-0.99	37	Westgate (1994)
P3978	282 ± 4.80	-5.8	500	-0.99	37	Westgate, unpublished
*P3978	281 ± 10.9	-6.0	514	-0.99	39	Westgate, unpublished
*P3978	221 ± 8.50	-4.9	431	-0.99	43	Westgate, unpublished
*P3978	187 ± 16.3	-3.9	341	-0.99	39	Westgate, unpublished
Wheat						
Sun9E	37.8 ± 1.0	-0.87	69	-0.99	36	Brooks, Jenner, and Aspinall (1982)
*Sun9E	29.2 ± 1.5	-0.76	61	-0.99	41	Brooks, Jenner, and Aspinall (1982)
Triple Dirk	57.2 ± 0.3	-1.30	107	-0.99	38	Sofield et al. (1977)
Stewart	49.7 ± 2.8	-1.35	102	-0.97	39	Saari, Salminem, and Hill (1985)
Rye						
Prolific	36.2 ± 1.3	-0.92	74	-0.96	41	Saari, Salminem, and Hill (1985)
Triticale						
6A190	52.2 ± 2.9	-0.94	90	-0.93	40	Saari, Salminem, and Hill (1985)
6A250	37.8 ± 3.2	-0.73	68	-0.92	41	Saari, Salminem, and Hill (1985)

Note: Max DW was estimated from linear regression analysis of kernel DW versus kernel moisture curves. *Indicates droughted during grain filling. Kernel moisture is calculated on a DW basis. Kernel DWs are shown as mean ± SD.

than do well-watered plants because they begin to desiccate earlier and reach this minimum moisture content for growth sooner after anthesis (Westgate, 1994).

This analysis strongly suggests that the water status of the kernel late in grain filling is an important determinant of the duration of kernel filling. How is the change in water status translated into a decrease in synthetic activity? The loss of water from the kernel during grain filling is not reflected in the water potential of the kernel until moisture content declines below about 350 grams per kilogram $(g \cdot kg^{-1})$ (Barlow et al., 1980; Westgate and Boyer, 1986b; Westgate, 1994). Beyond this point, grain water potential decreases rapidly with moisture content due to the development of large matric forces (Vertucci, 1989). Egli and TeKrony (1997) reported water potentials of maize and wheat kernels were generally between –1.6 and –2.0 megapascals (MPa) at physiological maturity (maximum dry weight). They suggested this low water potential might trigger processes that ultimately lead to a cessation of seed growth. Other studies on cereals, however, show that dry-matter accumulation ceases before large changes in kernel water potential (or osmotic potential) occur late in grain filling (Barlow et al., 1980; Westgate and Boyer, 1986b; Westgate, 1994). Given the rapid change in water potential as moisture contents decline, the difficulty in determining exactly when reserve deposition ceases, and the spatial variation in moisture content within the kernel, it is unlikely we can define a unique water potential at which reserve deposition ceases.

Extending the duration of filling in droughted plants requires a greater understanding of the factors that might limit reserve deposition in the kernels late in development. The rapid synthesis of end products such as starch, protein, and oil during grain filling requires optimum coordination between substrate availability, enzyme activation, translation, and transcription (Bewley and Black, 1994). Concentrations of sucrose and amino acids per gram (g) of water in the maize endosperms remain fairly constant during linear grain fill, then increase as the water content declines (Doehlert, 1990). The increase in solute concentration occurs well in advance of physiological maturity (Westgate and Boyer, 1986b; Westgate, 1994), which strongly suggests that end product formation was not limited by lack of substrate when dry-matter accumulation ceased. Likewise, activities of enzymes involved in nitrogen and carbohydrate metabolism decrease late in grain fill (Doehlert, Min Kou, and Felker, 1986; Singletary et al., 1990; Muhitch, 1991). In vitro studies show that desiccation inhibits enzyme activity dramatically, but only at very low moisture contents typical of *dry* seeds (Stevens and Stevens, 1977; Rupley,

Gratton, and Careri, 1983). Enzymes extracted from physiologically mature seeds retain fairly high levels of activity (Doehlert, Min Kou, and Felker, 1986; Muhitch, 1991). Although such in vitro measurements likely overestimate in vivo activity, they suggest that the osmotic conditions prevalent in the endosperm and embryo late in grain filling do not inhibit enzyme activity directly.

The loss of enzyme activity probably reflects a decreased capacity for protein synthesis as water content declines (Bewley, 1981). In wheat endosperms, for example, the decline in protein synthesis is closely coupled to the level of translatable mRNAs (messenger ribonucleic acids) (Greene, 1983; Kermode et al., 1986; Kermode, Oishi, and Bewley, 1989). Also, premature desiccation of caster bean (*Ricinus communis* L.) seeds caused the same quantitative and qualitative changes in developmental mRNAs that occur during normal maturation (Kermode, Oishi, and Bewley, 1989). Unfortunately, studies on mRNA production in maize kernels have not been extended to the later stages of grain filling (Viotta, Sala, and Soave, 1975; Jones, 1978; Marks, Lindell, and Larkins, 1985). Similar mRNA profiles in kernels of well-watered and water-deficient plants during tissue desiccation would support the hypothesis that drought shortens the duration of dry-matter accumulation by accelerating the programmed shift from transcription of synthetic mRNAs to those required for desiccation tolerance and germination (Dure, 1997).

CONCLUSIONS

- Recent evidence indicates that selection for a minimum anthesis-silking interval is an indirect selection for rapid ear growth prior to and during anthesis. A short, or even negative, ASI is coupled to rapid ear growth, which maintained sucrose flux to the ear and spikelet development under stress.
- Improvements in kernel set under drought conditions have been coupled closely with the maintenance of ovary growth, which depends on carbohydrate metabolism. Drought causes lesions in this metabolism within the developing ovaries, which diminished their capacity to utilize sucrose supplied by the plant. The molecular basis of this phenomenon is not yet known.

- The stability of kernel growth during Stage II implies that the rate of weight gain is determined genetically and is moderated by conditions within the kernel itself. Presumably, maintenance of a favorable water status within the kernel permits metabolism to continue despite severe water deficits in vegetative tissues.
- Maximum kernel water volume, established at the end of Stage I, could determine final kernel size by (1) setting an upper limit on cellular volume available for dry-matter accumulation and (2) starting the "developmental clock" for grain-filling duration. Grain filling apparently continues until metabolism is limited by low moisture content.

REFERENCES

Adams, C.A. and Rime, R.W. (1980). Moisture content as a controlling factor in seed development and germination. *International Review of Cytology* 68: 1-8.

Artlip, T.S., Madison, J.T., and Setter, T.L. (1995). Water deficit in developing endosperm of maize: Cell division and nuclear DNA endoreduplication. *Plant, Cell and Environment* 18: 1034-1040.

Barlow, E.W.R., Lee, J.W., Munns, R., and Smart, M.G. (1980). Water relations of the developing wheat grain. *Australian Journal of Plant Physiology* 7: 519-525.

Bassetti, P. and Westgate, M.E. (1994). Floral asynchrony and kernel set quantified by image analysis. *Agronomy Journal* 86: 699-703.

Bewley, J.D. (1981). Protein synthesis. In Paleg, L.G. and Aspinall, D. (Eds.), *The physiology and biochemistry of drought resistance in plants* (pp. 261-282). Sydney, Australia: Academic Press.

Bewley, J.D. and Black, M. (1994). *Seeds: Physiology of development and germination,* Second edition. New York: Plenum Press.

Bolaños, J. and Edmeades, G.O. (1993). Eight cycles of selection for drought tolerance in lowland tropical maize. II. Responses in reproductive behavior. *Field Crops Research* 31: 253-268.

Boyle, M.G., Boyer, J.S., and Morgan, P.W. (1991). Stem infusion of liquid culture medium prevents reproductive failure of maize at low water potentials. *Crop Science* 31: 1246-1252.

Bradford, K.J. (1994). Water stress and the water relations of seed development: A critical review. *Crop Science* 34: 1-11.

Brooks, A., Jenner, C.F., and Aspinall, D. (1982). Effects of water deficit on endosperm starch granules and on grain physiology of wheat and barley. *Australian Journal of Plant Physiology* 9: 423-436.

Cheikh, N. and Jones, R.J. (1994). Disruption of maize kernel growth and development by heat stress: Role of cytokinin abscisic acid balance. *Plant Physiology* 106: 45-51.

Cirilo, A.G. and Andrade, F.H. (1996). Sowing date and kernel weight in maize. *Crop Science* 36: 325-331.

Cobb, B.G., Hale, D.J., Smith, J.D., and Kent, M.W. (1988). The effects of modifying sucrose concentration on the development of maize kernels grown in vitro. *Annals of Botany* 62: 265-270.

Dietrich, J.T., Kaminek, M., Blevins, D.G., Reinbott, T.M., and Morris, R.O. (1995). Changes in cytokinins and cytokinin oxidase activity in developing maize kernels and the effects of exogenous cytokinin on kernel development. *Plant Physiology and Biochemistry* 33: 327-336.

Doehlert, D.C. (1990). Distribution of enzyme activities within the developing maize (*Zea mays*) kernel in relation to starch, oil, and protein accumulation. *Physiologia Plantarum* 78: 560-567.

Doehlert, D.C., Min Kou, T., and Felker, F.C. (1986). Enzymes of sucrose and hexose metabolism in developing kernels of two inbreds of maize. *Plant Physiology* 86: 1013-1019.

DuPlessis, D.P. and Dijkhuis, F.J. (1967). The influence of the time lag between pollen shedding and silking on the yield of maize. *South African Journal of Agricultural Science* 10: 667-674.

Dure, L.I. (1997). Lea proteins and the desiccation tolerance of seeds. In Larkins, B.A. and Vasil, I.K. (Eds.), *Cellular and molecular biology of plant seed development* (pp. 525-543). Dordrecht, Netherlands: Kluwer Academic Publishers.

Edmeades, G.O., Bolaños, J., Bänziger, M., Chapman, S., Ortega, A., Lafitte, H.R., Fischer, K.S., and Pandey, S. (1997). Recurrent selection under managed-drought stress improves grain yields in tropical maize. In Edmeades, G.O., Bänziger, M., Mickelson, H.R., and Peña-Valdivia, C.B. (Eds.), *Developing drought- and low-N-tolerant maize* (pp. 415-425). Mexico DF, Mexico: CIMMYT.

Edmeades, G.O., Bolaños, J., Hernandez, M., and Bello, S. (1993). Causes for silk delay in a lowland tropical maize population. *Crop Science* 33: 1029-1035.

Egli, D.B. (1994). Seed growth and development. In Boote, K.J., Bennett, J.M., Sinclair, T.R., and Paulsen, G.M. (Eds.), *Physiology and determination of crop yield* (pp. 127-148). Madison, WI: ASA-CSSA-SSSA.

Egli, D.B. (1998). *Seed biology and the yield of grain crops.* Cary, NC: Oxford University Press.

Egli, D.B. and TeKrony, D.M. (1997). Species differences in seed water status during seed maturation and germination. *Seed Science Research* 7: 3-11.

Fisher, D.B. and Gifford, R.M. (1986). Accumulation and conversion of sugars by developing wheat grains. VI. Gradients along the transport pathway from the peduncle to the endosperm cavity during grain filling. *Plant Physiology* 82: 1024-1030.

Frederick, J.R., Hesketh, J.D., Peters, D.B., and Below, F.E. (1989). Yield and reproductive trait responses of maize hybrids to drought stress. *Maydica* 34: 319-328.

Grant, R.F., Jackson, B.S., Kiniry, J.R., and Arkin, G.K. (1989). Water deficit timing effects on yield components on maize. *Agronomy Journal* 81: 61-65.

Greene, F.C. (1983). Expression of storage protein genes in developing wheat (*Triticum aestivum* L.) seeds. *Plant Physiology* 71: 40-46.

Jenner, C.F., Ugalde, T.D., and Aspinall, D. (1991). The physiology of starch and protein deposition in the endosperm of wheat. *Australian Journal of Plant Physiology* 18: 211-226.

Jones, R.A. (1978). Effect of floury-2 locus on zein accumulation and RNA metabolism during maize endosperm development. *Biochemical Genetics* 16: 27-38.

Jones, R.J. (1994). Intrinsic factors regulating seed development. In Boote, K.J., Bennett, J.M., Sinclair, T.R., and Paulson, G.M. (Eds.), *Physiology and determination of crop yield* (pp. 149-152). Madison, WI: ASA.

Jones, R.J., Roessler, J., and Ouattar, S. (1985). Thermal environment during endosperm cell division in maize: Effects on number of endosperm cells and starch granules. *Crop Science* 25: 830-834.

Jones, R.J., Schreiber, B.M.N., and Roessler, J.A. (1996). Kernel sink capacity in maize: Genotype and maternal regulation. *Crop Science* 36: 301-306.

Jones, R.J. and Simmons, R.R. (1983). Effect of altered source-sink ratio on growth of maize kernels. *Crop Science* 23: 129-134.

Jurgens, S.K., Johnson, R.R., and Boyer, J.S. (1978). Dry matter production and translocation in maize subjected to drought during grain fill. *Agronomy Journal* 70: 678-682.

Keeling, P.L., Banisadr, R., Barone, L., Wasserman, B.P., and Singletary, G.W. (1994). Effect of temperature on enzymes in the pathway of starch biosynthesis in developing wheat and maize grain. *Australian Journal of Plant Physiology* 21: 807-827.

Kermode, A.R., Bewley, D.J., Dasgupta, J., and Misra, S. (1986). The transition from seed development to germination: A key role for desiccation? *Hortscience* 21.

Kermode, A.R., Oishi, M.Y., and Bewley, J.D. (1989). Regulatory roles for desiccation and abscisic acid in seed development: A comparison of the evidence from whole seeds and isolated embryos. In Stanwood, P.C. and McDonald, M.B. (Eds.), *Seed moisture*, Volume 14 (pp. 23-50). Madison, WI: CSSA.

Koch, K.E., Nolte, K.D., Duke, E.R., McCarty, D.R., and Avigne, W.T. (1992). Sugar levels modulate differential expression of maize sucrose synthase genes. *Plant Cell* 4: 59-69.

Lur, H.S. and Setter, T.L. (1993). Endosperm development of maize defective kernel (dek) mutants. Auxin and cytokinin levels. *Annals of Botany* 72: 1-6.

Mambelli, S. and Setter, T.L. (1998). Inhibition of maize endosperm cell division and endoreduplication by exogenously applied abscisic acid. *Physiologia Plantarum* 104: 266-272.

Marks, M.D., Lindell, J.S., and Larkins, B.A. (1985). Quantitative analysis of the accumulation of zein mRNA during maize endosperm development. *Journal of Biological Chemistry* 260: 16445-16450.

Moser, S., Feil, B., Thiraporn, R., and Stamp, P. (1997). Tropical maize under pre-anthesis drought and low nitrogen supply. In Edmeades, G.O., Bänziger, M., Mickelson, H.R., and Peña-Valdivia, C.B. (Eds.), *Developing drought- and low-N-tolerance maize* (pp. 159-162). Mexico DF, Mexico: CIMMYT.

Muhitch, M.J. (1991). Tissue distribution and developmental patterns of NADH-dependent and ferridoxin-dependent glutamate synthase activities in maize (*Zea mays* L.) kernels. *Physiologia Plantarum* 81: 481-488.

NeSmith, D.S. and Ritchie, J.T. (1992). Maize (*Zea mays* L.) response to a severe soil water deficit during grain filling. *Field Crops Research* 29: 23-35.

Ouattar, S., Jones, R.J., and Crookston, R.K. (1987). Effect of water deficit during grain filling on the pattern of maize kernel growth and development. *Crop Science* 27: 726-730.

Ouattar, S., Jones, R.J., Crookston, R.K., and Kajeiou, M. (1987). Effect of drought on water relations of developing maize kernels. *Crop Science* 27: 730-735.

Passioura, J.B. (1994). The yield of crops in relation to drought. In Boote, K.J., Bennett, J.M., Sinclair, T.R., and Paulson, G.M. (Eds.), *Physiology and determination of crop yield* (pp. 343-360). Madison, WI: ASA.

Porter, G.A., Kneivel, D.P., and Shannon, J.C. (1985). Sugar efflux from maize (*Zea mays* L.) pedicel tissue. *Plant Physiology* 77: 524-531.

Reed, A.J. and Singletary, G.W. (1989). Roles of carbohydrate supply and phytohormones in maize kernel abortion. *Plant Physiology* 91: 986-992.

Rupley, J.A., Gratton, E., and Careri, G. (1983). Water and globular proteins. *Trends in Biochemistry* 8: 18-22.

Saari, L., Salminem, S.O., and Hill, R.D. (1985). Sucrose and sucrose synthase activity during kernel development in triticale, wheat, and rye. *Canadian Journal of Plant Science* 65: 867-877.

Saini, H.S. (1997). Effects of water stress on male gametophyte development in plants. *Sex Plant Reproduction* 10: 67-73.

Saini, H.S. and Westgate, M.E. (2000). Reproductive development in grain crops during drought. *Advances in Agronomy* 68: 59-96.

Schussler, J.R. and Westgate, M.E. (1991a). Maize kernel set at low water potential. I. Sensitivity to reduced assimilates during early kernel growth. *Crop Science* 31: 1189-1195.

Schussler, J.R. and Westgate, M.E. (1991b). Maize kernel set at low water potential. II. Sensitivity to reduced assimilates at pollination. *Crop Science* 31: 1196-1203.

Schussler, J.R. and Westgate, M.E. (1994). Increasing assimilate reserves does not prevent kernel abortion at low water potential in maize. *Crop Science* 34: 1569-1576.

Schussler, J.R. and Westgate, M.E. (1995). Assimilate flux determines kernel set at low water potential in maize. *Crop Science* 35: 1074-1080.

Singletary, G.W., Doehlert, D.C., Wilson, C.M., Muhitch, M.J., and Below, F.E. (1990). Response of enzymes and storage proteins of maize endosperm to nitrogen supply. *Physiologia Plantarum* 94: 858-864.

Sofield, I., Wardlaw, I.F., Evans, L.T., and Zee, S.Y. (1977). Nitrogen, phosphorus and water contents during grain development and maturation in wheat. *Australian Journal of Plant Physiology* 4: 799-810.

Stevens, E. and Stevens, L. (1977). Glucose-6-phosphate dehydrogenase activity under conditions of water limitation: A possible model system for enzyme reaction in unimbibed resting seeds and its relevance to seed viability. *Journal of Experimental Botany* 28: 292-303.

TeKrony, D.M. and Hunter, J.L. (1995). Effect of seed maturation and genotype on seed vigor in maize. *Crop Science* 35: 857-862.

Vertucci, C.M.P. (1989). The effect of low water contents on physiological activities of seeds. *Physiologia Plantarum* 77: 172-176.

Viotta, A., Sala, E., and Soave, C. (1975). RNA metabolism and polysome profiles during seed development in normal and opaque-2 maize endosperms. *Maydica* 20: 111-124.

Wendler, R., Veith, R., Dancer, J., Stitt, M., and Komor, E. (1991). Sucrose storage in cell suspension cultures of *Saccharum* sp. (sugarcane) is regulated by a cycle of synthesis and degradation. *Planta* 183: 31-39.

Westgate, M.E. (1994). Water status and development of the maize endosperm and embryo during drought. *Crop Science* 34: 76-83.

Westgate, M.E. and Boyer, J.S. (1985). Carbohydrate reserves and reproductive development at low leaf water potentials in maize. *Crop Science* 25: 762-769.

Westgate, M.E. and Boyer, J.S. (1986a). Reproduction at low silk and pollen water potentials in maize. *Crop Science* 26: 951-956.

Westgate, M.E. and Boyer, J.S. (1986b). Water status of the developing grain of maize. *Agronomy Journal* 78: 714-719.

Westgate, M.E. and Grant, D.T. (1989). Water deficits and reproductive development in maize. Response of the reproductive tissues to water deficits in anthesis and mid-grain fill. *Plant Physiology* 91: 862-867.

Zinselmeier, C. (1991). The role of assimilate supply, partitioning, and metabolism in maize kernel development at low water potential. PhD Thesis, University of Minnesota, St. Paul.

Zinselmeier, C., Lauer, M.J., and Boyer, J.S. (1995). Reversing drought-induced losses in grain yield: Sucrose maintains embryo growth in maize. *Crop Science* 35: 1390-1400.

Zinselmeier, C., Schussler, J.R., Westgate, M.E., and Jones, R.J. (1995). Low water potential disrupts carbohydrate metabolism in maize ovaries. *Plant Physiology* 107: 385-391.

Zinselmeier, C., Westgate, M.E., and Jones, R.J. (1995). Kernel set at low water potential does not vary with source/sink ratio in maize. *Crop Science* 35: 158-163.

Chapter 8

Variation in Apical Dominance and Its Implications for Herbivory Resistance, Competitive Ability, and Biomass Partitioning

Víctor O. Sadras

INTRODUCTION

Apical dominance is the control exerted by the apical portions of the shoot over the outgrowth of the lateral buds (Cline, 1991). In common with many other crop species, domestication and breeding for yield and ease of harvest in maize favored plant types with strong apical dominance, simplified architecture, and less plastic growth forms (Evans, 1993). These processes led from ancestral types able to produce numerous tillers and ears to modern uniculm forms that produce one or few ears (Doebley, Stec, and Hubbard, 1997; Rosenthal and Welter, 1995).

Three aspects of prolificacy, and the underlying phenomenon of apical dominance, are important in maize: (1) its role in the adaptation of the crop to a range of environments, (2) its responsiveness to environmental factors and crop management, and (3) its variation among genotypes. In a comparison among maize, sunflower, and soybean, Vega (1997) plotted grain number versus plant growth rate around flowering. Plant growth rate, in turn, changed as a function of plant population density. In all three species, grain number increased

The author would like to thank C. R. C. Vega for stimulating discussion, and gratefully acknowledge the financial support received from Fundación Antorchas (Grant A-13388/1-2).

almost linearly with plant growth rate up to about 5 grams per day (g/day). After this threshold there was a clear decline in the response of sunflower grain number, reflecting the sharp limit imposed by the single infrutescence in commercial cultivars of this species. The response of maize resulted from the overlap of two "sunflower-like" response curves, highlighting the importance of secondary ears in the adjustment to increasing availability of resources (see Figure 8.1). Vega (1997) proposed that grain number in indeterminate soybean does not saturate because additional reproductive sinks are generated with increasing availability of resources. Empirical support for this statement is limited because of practical problems in extending the

FIGURE 8.1. Relationship Between Kernel Number per Plant and Plant Growth Rate During the Critical Period of Kernel Set in Maize (cv DK636)

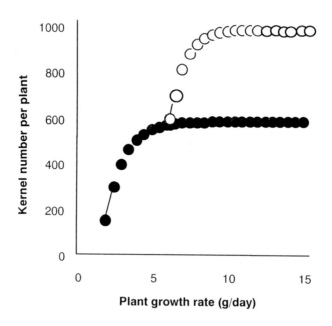

Source: Adapted from Vega (1997).

Note: Closed symbols: first ear; open symbols: second ear.

range of soybean plant growth rate beyond 5 g/day (C. Vega, personal communication). Variation in the degree of apical dominance and associated growth strategies thus account for part of the differences among the three crop species.

The variation in the degree of apical dominance among and within species can be addressed from different perspectives. From a physiological viewpoint, variation in apical dominance can be analyzed in terms of nutritional, hydraulic, and hormonal factors (Cline, 1991, 1994; McIntyre, 1997; Töpperwein, 1993).

From a complementary ecological perspective, variation in apical dominance can be analyzed for its effects on plant functionality and fitness (Aarssen, 1995; Bonser and Aarssen, 1996). About half of all known insect species are more or less dependent on vascular plants (Panda and Khush, 1995), and the interaction between herbivores and their hosts has been a major factor in the selection of certain plant traits (Barnes, 1989). Thus, from an ecological perspective, mathematical models have been developed that predict the degree of bud dormancy—directly related to degree of apical dominance—in relation to the intensity and risk of herbivory (Nilsson, Tuomi, and Åström, 1996; Tuomi, Nilsson, and Åström, 1994). The relations between apical dominance and herbivory tolerance have also been investigated experimentally (Paige and Whitham, 1987; Sadras and Fitt, 1997). Aarssen (1995) postulated a series of hypotheses to account for the evolution of apical dominance in plants and emphasized the role of this trait in relation to pollination, seed dispersal, herbivory, and competition. Intraspecific variation in apical dominance is widespread, as has been reported (see Table 8.1) for plants from different habitats (e.g., tropical rainforests to aridlands), life histories (perennial to annual), and agronomic selection (wild to cultivated). Bonser and Aarssen (1996) proposed a classification theory for adaptive strategies in herbaceous plants based on the allocation of meristems.

From an ecological viewpoint, this chapter highlights the variation in apical dominance in a range of cultivated and wild species and discusses the putative effects of varying apical dominance upon herbivory resistance and competitive ability. The reproductive responses of soybean, maize, and sunflower to growing conditions outlined previously are briefly reanalyzed using the meristem allocation framework of Bonser and Aarssen (1996).

TABLE 8.1. Empirical Evidence for Intraspecific Variation in Apical Dominance

Species	Assessment of apical dominance	Reference
Cedrela odorata	Response to decapitation	(Newton et al., 1995)
Daphne laureola	Branching	(Alonso and Herrera, 1996)
Gossypium spp.	Branching/response to decapitation	(Sadras and Fitt, 1997)
Triplochiton scleroxylon	Response to decapitation	(Ladipo, Leaky, and Grace, 1991)
Triticum aestivum	Tillering and height	(Huel and Hucl, 1996)
Zea mays	Prolificacy	(Motto and Moll, 1983)
Zea spp. (maize-teosinte)	Tillering, prolificacy	(Rosenthal and Welter, 1995)

APICAL DOMINANCE AND HERBIVORY RESISTANCE

This section illustrates how variation in apical dominance may affect resistance to herbivory in several plant-herbivore systems.* Theoretical as well as empirical evidence is considered. The responses of architecturally diverse *Zea* spp. to the stem borer *Diatraea grandiosella* are analyzed in some detail.

Theory

Van der Meijden (1990) compares two plant types: *a,* with restrained stem growth, followed by triggering of accelerated stem growth after herbivory, and *b,* which invests all available resources in growth, and proposes a simple model of the Malthusian fitness parameter (λ), or the expected rate of increase of these plant types, as follows:

---

*Plant reactions to a given herbivore grade, continuously from full resistance to the extreme sensitivity of plants that are unprotected and unable to regrow after damage. Herbivory resistance involves two components: avoidance and tolerance (Rosenthal and Kotanen, 1994). Plants can avoid damage via phenological escape or defenses. Tolerance, a complex trait, is the capacity of a plant to withstand herbivory with minimal losses in growth and reproduction.

$$\lambda_a = H \, S_h + (1-H) \, S_a \qquad (8.1)$$

$$\lambda_b = (1-H) \, S_b \qquad (8.2)$$

where H is the proportion of plants that suffer herbivory; S_a and S_h are seed production of plant type *a* before and after herbivory, respectively; and S_b is seed production of un-damaged plant type *b*.

The model assumes that damaged plants *b* die because the lack of dormant meristems precludes their recovery after damage. Further considerations, supported by empirical values of S_a, S_h, and S_b measured by Paige and Whitham (1987) in a system of scarlet gilia (*Ipomopsis aggregata*) browsed by ungulates, lead to the conclusion that λ_a is greater than λ_b if H is greater than 0.5. Some degree of apical dominance, concludes van der Meijden (1990), would be favored in habitats where the risk of herbivory is greater than 50 percent, herbivory is restricted to a limited period, and the timing of herbivory is unpredictable from other environmental clues. More elaborate models predict optimum ranges of bud dormancy as a function of risk of herbivory and activation sensitivity of dormant buds in relation to meristem loss (see Table 8.2).

Observation

Substantial evidence indicates that release of apical dominance following herbivory and subsequent growth from axillary meristems is a critical mechanism of tolerance or "compensation" in many plant species (Moore, 1994; Sadras, 1996; Sadras, 1997; Wein and Minotti, 1988) (see Table 8.2). Variation in apical dominance and plant architecture can thus strongly influence the outcome of the interaction between plants and their herbivores. Other more subtle links between apical dominance and herbivory can be mediated by variation in both vascular connectivity and plant tissue quality as food for herbivores.

Paige and Whitham (1987) showed that release of apical dominance in heavily browsed scarlet gilia triggered a twofold increase in fitness—seed production and subsequent seedling survival—in relation to uneaten controls. Further studies involving measures of plant

TABLE 8.2. Empirical and Theoretical Evidence for the Association Between Apical Dominance and Herbivory Resistance

Plant species	Method	Observation	Reference
Betula pubescens spp. *tortuosa*	Simulated damage	Disturbances of apical dominance change the quality of plant tissues as food for herbivores.	(Senn and Haukioja, 1994)
Generic	Mathematical models	Varying apical dominance maximizes fitness (seed production) depending on the risk of herbivory.	(Nilsson, Tuomi, and Åström, 1996; Tuomi, Nilsson, and Åström, 1994; van der Meijden, 1990)
Gossypium hirsutum	Simulated or actual insect herbivory	Release of apical dominance after damage allows for plant recovery and, in some cases, "overcompensation."	(Danobrega et al., 1993; Evenson, 1969; Heilman, Nakmen, and Dilday, 1981; Sadras and Fitt, 1997)
Ipomea aggregata	Mammalian herbivory	High apical dominance decreased seed production.	(Paige and Whitham, 1987)
Zea spp.	Response to stemborer damage	Herbivory tolerance is (stem borer) associated with low apical dominance and enhanced tiller production in response to damage.	(Rosenthal and Welter, 1995)

death in populations inside and outside deer exclosures showed that herbivory is not necessarily beneficial to scarlet gilia populations, as formerly concluded by Paige and Whitham (Bergelson and Crawley, 1992a, b).

In a system comprising twelve to nineteen cotton varieties, *Helicoverpa* spp. (Lepidoptera: Noctuidae), and sucking mirid bugs (Hemiptera: Miridae), plant resistance to herbivory was quantified as the seed yield ratio between unprotected and insecticide-protected crops (Sadras and Fitt, 1997). Resistance was nonlinearly related to apical dominance, with a broad optimum and declining resistance after a threshold of apical dominance was surpassed. Three reasons were proposed to account for the poor resistance of plants with very strong apical dominance. First, they were taller and, therefore, more visible and attractive to insects. Second, genotypes with strong apical dominance require more time for regrowth, which depends on activation of axillary buds, and this could lead to less time available for recovery. Third, activation of axillary buds favors not only vegetative but also reproductive growth.

In systems of *Daphne laureola* (Thymelaeaceae) and noctuid larvae (Lepidoptera: Noctuidae), variation in plant architecture was associated with important differences in the incidence of larvae among, as well as within, plants (Alonso and Herrera, 1996). Variation among plants in the incidence of larvae was directly related to the number of leaf whorls, and inversely to the mean basal diameter of stems. This response was attributed to discrimination by ovipositing females. Within plants, larvae preferentially selected leaf whorls with shorter supporting stems and lower branching orders. This was attributed to movement costs, which may have been reduced by larvae using plant architectural traits as cues for within-plant food selection.

Plant architecture and its associated vascular connections between damaged and undamaged plant parts provide the framework for the integration, at the plant level, of wound-induced chemical responses—although some studies failed to find relations between vascular architecture and herbivore-induced resistance (Mutikainen, Walls, and Ojala, 1996). Senn and Haukioja (1994) concluded that the sensitivity of mountain birch, *Betula pubescens,* to disturbances in apical dominance might make it possible for herbivores to manipulate the quality of their food source.

In conclusion, theoretical and empirical studies strongly support the notion that variation in apical dominance has substantial effects on herbivory resistance in both wild and cultivated plants.

Resistance to Stem Borer in Architecturally Diverse Zea spp.

Rosenthal and Welter (1995) demonstrated that, in common with other species such as alfalfa (Small, 1996), domestication of maize reduced the effectiveness of several natural adaptations against insect herbivores. Part of the increased susceptibility to insects is related to changes in apical dominance and plant architecture.

Rosenthal and Welter (1995) investigated the resistance to larvae of the stem borer *Diatraea grandiosella* (Pyralidae) in plants of the maize-teosinte complex, including the wild perennial *Zea diploperennis;* the wild annual teosinte, *Z. mays* ssp. *parviglumis;* a landrace cultivar of maize, Reventador, and a modern high-yielding variety, VS524 (both *Z. mays* ssp. *mays*). These are contrasting plants in life history (perennial versus annual), domestication (wild versus cultivated), and agronomic selection (landrace versus modern variety) (see Table 8.3).

A marked gradient of architectural complexity is evident from the very complex wild perennial to the much simpler modern cultivar. Tiller number ranges from 11 to 1, and leaf number, from 71 to 14 (see Table 8.3). The consistent reduction in tillering from the wild perennial to the cultivated types can be attributed, at least partially, to stronger apical dominance. Variation in apical dominance is also reflected in differences in plant height between the wild perennial (1.5 meters [m]) and the other plants (2.2 m). Interestingly, most of the differences in plant architecture between teosinte and maize are accounted for by a small number of quantitative trait loci (QTL) (Doebley and Stec, 1993). Doebley, Stec, and Hubbard (1997) summarized the large effects of the QTL corresponding to the teosinte branched 1 (*tb1*) locus. The effects of *tb1* alone are sufficiently large that teosinte fully converts to maize plant architecture when carrying this gene. On the other hand, the *tb1* mutation in maize causes a complete loss of apical dominance.

Both components of herbivory resistance, avoidance and tolerance, varied between plant types. Defense, a major component of avoidance, was greatest in the wild perennial, followed by the annual wild relative, and least in the domesticated plants (see Table 8.3). Resistance, quantified as dry matter or grain yield reduction due to stem borer relative to undamaged controls, was greatest in *Zea diploperennis* (see Table 8.3). This was partially associated with variation in tolerance, (i.e., the architectural complexity of *Z. diploperennis* allowed for compartmentalization of damage and greater developmental plasticity). Resistance, quantified as a function of plant height, and apical dominance were inversely related (see Table 8.3).

TABLE 8.3. Comparison of *Zea* spp. Spanning a Range of Architectures, Degree of Domestication, and Herbivory Resistance

	Species			
Characteristic	*Z. diploperennis*	*Z. mays* ssp. *parviglumis*	*Z. mays* ssp. *mays* cv. Reventador	*Z. mays* ssp. *mays* cv. VS524
Life history	perennial	annual	annual	annual
Domestication	wild	wild	domesticated (landrace)	domesticated (high-yielding cultivar)
Tiller number [1]	11	7	3	1
Leaf number [1]	71	57	26	14
Apical dominance[1]	+	+	++	+++
Tillering response to herbivory	increased	none	none	none
Defense level	+++	++	+	+
Resistance = *f* (biomass)	+++	+	+	+
f (grain yield)	+++	+	not measured	+
f (height)	+++	++	+	+

Source: Adapted from Rosenthal and Welter (1995).

Note: +++, ++, and + indicate greatest, intermediate, and lowest, respectively.

[1] Assessed in undamaged plants

It is interesting to note that the wild perennial was the only plant that increased tillering in response to stem borer damage. A comparable range of tillering responses can be observed in wheat after damage by Hessian fly (*Mayetiola destructor:* Cecidomyiidae) (Sadras, Fereres, and Ratcliffe, 1999). Wellso and colleagues (Wellso and Araya, 1993; Wellso and Hoxie, 1994) showed that infested wheat plants may compensate for Hessian fly injury by enhanced tiller production if the infested main stem is not killed, and they also demonstrated the existence of intraspecific variability in tillering responses to Hessian fly.

In summary, domestication and selection for yield and harvestability increased maize apical dominance and decreased two major associated traits: tillering and prolificacy. In parallel to these changes, defenses and tolerance to *Diatraea grandiosella* decreased. The study of Rosenthal and Welter supports the existence of a causal link between reduced apical dominance and reduced tolerance to the stem borer.

APICAL DOMINANCE AND COMPETITIVE ABILITY

Aarssen (1995) pointed out that the most obvious advantage of vertical growth in plants is associated with competitive ability for light—a statement supported by observations in natural systems in which dominant species are commonly those capable of reaching the greatest heights.

The association between competitive ability and plant height has also been demonstrated in cultivated species. This is illustrated by Huel and Hucl (1996) in a comparison of sixteen wheat genotypes grown in competition with oat (*Avena sativa*) or oriental mustard (*Brassica juncea*) (see Figure 8.2). Mustard and oat yielded more when grown with shorter, less competitive wheat varieties, which, in turn, had greater yield reductions. Wheat height accounted for 62 to 80 percent of the variation in mustard or oat yield and for 58 percent of the variation in wheat yield reduction (see Figure 8.2).

Given the positive correlation between degree of apical dominance and plant height (Sadras and Fitt, 1997), and the impact of plant height on competitive ability, strong apical dominance should be favored by natural selection in "K type" habitats where resources abound and plants are normally crowded (Aarssen, 1995).

APICAL DOMINANCE, HERBIVORY RESISTANCE, AND COMPETITIVE ABILITY

Aarssen (1995) discussed the effect of variation in apical dominance on seed dispersion, pollination, foraging resources in poor habitats, competition for light in resource-rich habitats, and herbivory resistance. Strong evidence indicates that some degree of apical dominance optimizes resistance to prevalent herbivory, while both plant height and

FIGURE 8.2. Plant Height and Competitive Ability in an Experimental System of Wheat and Weeds (Mustard and Oats)

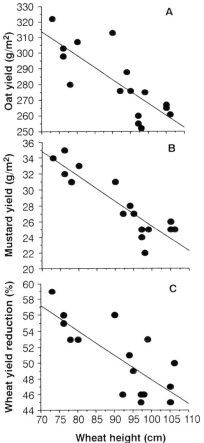

Source: Adapted from Huel and Hucl (1996).

Note: (A) Oat and (B) mustard yield decreased with increasing wheat height, while (C) wheat yield reduction decreased with wheat height. Experiments compared sixteen wheat varieties grown in different conditions of interspecific competition, i.e., no competition, moderate (oat or mustard sown at 48 seeds/m^2), or high competition (oat or mustard sown at 96 seeds/m^2); in all cases, wheat sowing rate was 180 seeds/m^2. Mustard and oat yields are averages of two sowing rates, and wheat yield reduction is the variation between yield of "pure" wheat and the yield of wheat averaged across two intensities of competition. The association between variables was significant in all cases ($P < 0.001$).

competitive ability for light increase with increasing degree of apical dominance. Considering these two major forces in the evolution of plants, it is speculated that a trade-off might be expected in environments where both herbivory and competition operate simultaneously (see Figure 8.3). The combined selective pressure of these factors should favor increasing apical dominance up to a point, indicated by the vertical line in Figure 8.3. After this threshold, further increase in apical dominance could increase competitive ability at the cost of reduced resistance to herbivory. Taller plants, for instance, could be more competitive but would also be more attractive to insects. The decline in herbivory resistance with increasing apical dominance in maize has been demonstrated by Rosenthal and Welter (1995).

Sadras and Fitt (1997) discuss empirical evidence that supports the existence of a trade-off between herbivory resistance and competitive ability that is mediated by variation in apical dominance (Blossey and Nötzold, 1995; Hartvigsen and McNaughton, 1995). Interestingly, variation in apical dominance can also underlie the negative association between trampling tolerance and competitive ability (Ikeda and Okutomi, 1995) and between resistance to grazing and flooding tolerance (Oesterheld and McNaughton, 1991).

The *tb1* mutants of maize described by Doebley, Stec, and Hubbard (1997) provide a system to test this hypothesis further; that is, in a mixture of wild type and *tb1* mutant, we could expect (1) the wild type to become dominant in the absence of herbivores and (2) the mutant to prevail in the presence of stem borers.

MERISTEM ALLOCATION:
A TOOL TO ANALYZE CROP GROWTH

The pattern of biomass partitioning is a major determinant of plant performance in both wild and cultivated species. In natural systems, plant strategies are often analyzed as a function of biomass partitioning, following some variation of the original model of Grime (1974). In cropping systems, the partitioning of biomass between reproductive and vegetative plant components is central to the management of key resources such as water (Cooper et al., 1987; Richards and Passioura, 1989; Sadras and Connor, 1991). In many crop species, enhancement of yield potential brought about by breeding and selec-

FIGURE 8.3. Relationship Between Apical Dominance and (A) Herbivory
Resistance and (B) Competitive Ability for Light

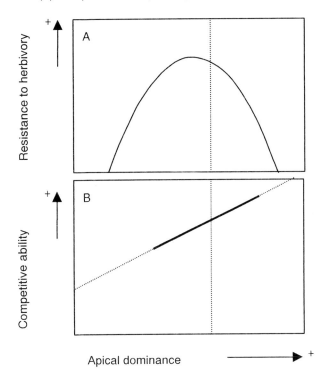

Source: Adapted from empirical studies with cotton (Sadras and Fitt, 1997).

Note: Solid lines indicate the actual range of measurements, while dashed lines
show extrapolation using fitted curves. The vertical line shows the threshold after
which a trade-off could be expected, whereby further increase in apical dominance
could increase competitive ability at the cost of reduced resistance to herbivory.

tion was associated with improved partitioning to harvestable organs
rather than to increased production of biomass (Slafer, 1994).

The pattern of meristem fate imposes clear limits on biomass
partitioning. Based on the concept of meristem allocation, Bonser
and Aarssen (1996) developed a classification of adaptive strategies
in herbaceous plants as an alternative to the conventional approach

based on the partitioning of biomass or other plant resources. They consider that all initially dormant axillary meristems have one of three principal fates: growth (G) meristems produce a new shoot or branch; reproductive (R) meristems produce a flower or inflorescence; or inactive (I) meristems remain dormant. Then, they propose to define plant strategies according to the measure of *reproductive effort* (RE), which is calculated as the percentage of R meristems, or $R/(G + I)$; *branching intensity,* $BI = G/(I + R)$; and *apical dominance,* $AD = I/(G + R)$.

The original model of Bonser and Aarssen (1996) aimed at ecological and evolutionary research. Nonetheless, their concepts may also be useful in a context of crop management and breeding. We could reanalyze, for instance, our initial comparison of maize, sunflower, and soybean using this meristem allocation approach. It is clear that the response of soybean to increasing availability of resources is the product of a strategy based on low AD, high RE, and high BI. In contrast, the low plasticity of maize and sunflower is the result of an extreme strategy with very strong AD, very low or nil BI, and low RE. This approach is useful not only in the characterization of intra- and interspecific variation but also to analyze plant responses to environmental factors. This is important in relation to the actual level of apical dominance that results from the effects of genotype, environment, and genotype × environment interactions. A closer look at the maize plants grown at low planting density in the work of Vega (1997) (see Figure 8.1, plant growth rate > 5/day) reveals interesting aspects of meristem allocation that can be interpreted within this frame (see Table 8.4).

About 19 percent of the plants in the crop produced tillers in response to favorable growing conditions. It is clear that this early commitment of meristems to vegetative growth somehow interfered with the development of a second ear, as only 2.5 percent of plants were able to produce both tillers and two ears, in contrast to 53 percent of no-tillering plants that produced a second viable ear.

CONCLUSION

The degree of apical dominance is a highly variable plant trait that strongly influences plant performance. This is because key plant

TABLE 8.4. Allocation of Meristems in Maize Plants Grown with High Availability of Resources.

	One ear	Two ears
Tillering plants	16.0	2.5
No-tillering plants	28.4	53.1

Source: Adapted from Vega (1997).

Note: Four categories were defined according to tillering and ear number. Values are percent of plants in the population within each category.

traits, including competitive ability and herbivory resistance, are closely related to the degree of apical dominance. It is speculated that trade-offs between major selective forces (e.g., herbivory and competition) might underlie the large and widespread intraspecific variation in apical dominance. Irrespective of the causes, apical dominance varies among genotypes and is an important trait to be considered in breeding and management of crops because allocation of meristems *precedes* and *frames* the pattern of biomass allocation. Analysis of meristem fates in terms of reproductive effort, branching intensity, and apical dominance could contribute to our understanding of plant biomass partitioning.

REFERENCES

Aarssen, L.W. (1995). Hypotheses for the evolution of apical dominance in plants: Implications for the interpretation of overcompensation. *Oikos* 74: 149-156.

Alonso, C. and Herrera, C.M. (1996). Variation in herbivory within and among plants of *Daphne laureola* (Thymelaeaceae): Correlation with plant size and architecture. *Journal of Ecology* 84: 495-502.

Barnes, R.D. (1989). *Invertebrate zoology.* New York: Saunders College Publishing.

Bergelson, J. and Crawley, M.J. (1992a). The effects of grazers on the performance of individuals and populations of scarlet gilia, *Ipomopsis aggregata. Oecologia* 90: 435-444.

Bergelson, J. and Crawley, M.J. (1992b). Herbivory and *Ipomopsis aggregata:* The disadvantages of being eaten. *American Naturalist.* 139: 870-882.

Blossey, B. and Nötzold, R. (1995). Evolution of increased competitive ability in invasive nonindigenous plants: A hypothesis. *Journal of Ecology* 83: 887-889.

Bonser, S.P. and Aarssen, L.W. (1996). Meristem allocation: A new classification theory for adaptive strategies in herbaceous plants. *Oikos* 77: 347-352.

Cline, M.G. (1991). Apical dominance. *Botanical Review* 57: 318-358.

Cline, M.G. (1994). The role of hormones in apical dominance. New approaches to an old problem in plant development. *Physiologia Plantarum* 90: 230-237.

Cooper, P.J.M., Gregory, P.J., Tully, D., and Harris, H.C. (1987). Improving water use efficiency of annual crops in rainfed systems of west Asia and North Africa. *Experimental Agriculture* 23: 113-158.

Danobrega, L.B., Beltrao, N.E.D., Vieira, D.J., Diniz, M.D., and Deazevedo, D.M.P. (1993). Effect of plant spacing and of apical bud removal period on herbaceous cotton. *Pesquisa Agropecuaria Brasileira* 28: 1379-1384.

Doebley, J. and Stec, A. (1993). Inheritance of the morphological differences between maize and teosinte: Comparison of results for two F2 populations. *Genetics* 134: 559-570.

Doebley, J., Stec, A., and Hubbard, L. (1997). The evolution of apical dominance in maize. *Nature* 386: 485-488.

Evans, L.T. (1993). *Crop evolution, adaptation and yield*. Cambridge: Cambridge University Press.

Evenson, J. (1969). Effects of floral and terminal bud removal on the yield and structure of the cotton plant in the Ord Valley, North Western Australia. *Cotton Growing Review* 46: 37-44.

Grime, J. (1974). Vegetation classification by reference to strategies. *Nature* 250: 26-31.

Hartvigsen, G. and McNaughton, S.J. (1995). Tradeoff between height and relative growth rate in a dominant grass from the Serengeti ecosystem. *Oecologia* 102: 273-276.

Heilman, M.D., Nakmen, L.N., and Dilday, R.H. (1981). Tobacco budworm: Effect of early-season terminal damage on cotton lint yield and earliness. *Journal of Economic Entomology* 74: 732-735.

Huel, D.G. and Hucl, P. (1996). Genotypic variation for competitive ability in spring wheat. *Plant Breeding* 115: 325-329.

Ikeda, H. and Okutomi, K. (1995). Effects of trampling and competition on plant growth and shoot morphology of *Plantago, Eragrostis* and *Eleusine* species. *Acta Botanica Neerlandica* 44: 151-160.

Ladipo, D.O., Leakey, R.R.B., and Grace, J. (1991). Clonal variation in apical dominance in *Triplochiton scleroxylon* K. Schum in response to decapitation. *Silvae Genetica* 40: 135-140.

McIntyre, G.I. (1997). The role of nitrate in the osmotic and nutritional control of plant development. *Australian Journal of Plant Physiology* 24: 103-118.

Moore, P.P. (1994). Yield compensation of red raspberry following primary bud removal. *Hortscience* 29: 701.

Motto, M. and Moll, R.H. (1983). Prolificacy in maize: A review. *Maydica* 28: 53-76.

Mutikainen, P., Walls, M., and Ojala, A. (1996). Herbivory-induced resistance in Betula pendula: The role of plant vascular architecture. *Oecologia* 108: 723-727.

Newton, A.C., Cornelius, J.P., Mesén, J.F., and Leakey, R.R.B. (1995). Genetic variation in apical dominance of *Cedrela odorata* seedlings in response to decapitation. *Silvae Genetica* 44: 2-3.

Nilsson, P., Tuomi, J., and Åström, M. (1996). Bud dormancy as a bet-hedging strategy. *American Naturalist* 147: 269-281.

Oesterheld, M. and McNaughton, S.J. (1991). Interactive effect of flooding and grazing on the growth of Serengeti grasses. *Oecologia* 88: 153-156.

Paige, K.N. and Witham, T.G. (1987). Overcompensation in response to mammalian herbivory: The advantage of being eaten. *American Naturalist* 129: 407-416.

Panda, N. and Khush, G.S. (1995). *Host plant resistance to insects*. Wallingford, UK: CAB International.

Richards, R.A. and Passioura, J.B. (1989). A breeding program to reduce the diameter of the major xylem vessel in the seminal roots of wheat and its effects on grain yield in rain-fed environments. *Australian Journal of Agricultural Research* 40: 943-950.

Rosenthal, J.P. and Kotanen, P.M. (1994). Terrestrial plant tolerance to herbivory. *Trends in Ecology and Evolution* 9: 145-148.

Rosenthal, J.P. and Welter, S.C. (1995). Tolerance to herbivory by a stemboring caterpillar in architecturally distinct maizes and wild relatives. *Oecologia* 102: 146-155.

Sadras, V.O. (1996). Population-level compensation after loss of vegetative buds: Interactions among damaged and undamaged cotton neighbours. *Oecologia* 106: 432-439.

Sadras, V.O. (1997). Effects of simulated insect damage and weed interference on cotton growth and reproduction. *Annals of Applied Biology* 130: 271-281.

Sadras, V.O. and Connor, D.J. (1991). Physiological basis of the response of harvest index to the fraction of water transpired after anthesis. A simple model to estimate harvest index for determinate species. *Field Crops Research* 26: 227-239.

Sadras, V.O., Fereres, A., and Ratcliffe, R.H. (1999). Wheat growth, yield and quality as affected by insect herbivores. In Satorre, E.H. and Slafer, G.A. (Eds.), *Wheat: Ecology and physiology of yield determination* (pp. 183-227). Binghamton, NY: Food Products Press.

Sadras, V.O. and Fitt, G.P. (1997). Apical dominance: Variability among *Gossypium* genotypes and its association with resistance to insect herbivory. *Environmental and Experimental Botany* 38: 145-153.

Senn, J. and Haukioja, E. (1994). Reactions of the mountain birch to bud removal—Effects of severity and timing, and implications for herbivores. *Functional Ecology* 8: 494-501.

Slafer, G.A. (1994). *Genetic improvement of field crops*. New York: Marcel Dekker, Inc.

Small, E. (1996). Adaptations to herbivory in alfalfa, *Medicago sativa. Canadian Journal of Botany* 74: 807-822.

Töpperwein, H. (1993). Relationships in the apical region of angiosperms. *Vereinigung für Angewandte Botanik* 67: 22-30.

Tuomi, J., Nilsson, P., and Åström, M. (1994). Plant compensatory responses: Bud dormancy as an adaptation to herbivory. *Ecology* 75: 1429-1436.

van der Meijden, E. (1990). Herbivory as a trigger for growth. *Functional Ecology* 4: 597-598.

Vega, C. (1997). Número de granos por planta en soja, girasol y maíz en función de las tasas de crecimiento por planta durante el período crítico de determinación del rendimiento. MSc thesis, University of Mar del Plata, Buenos Aires, Argentina, 45 pp.

Wein, H.C. and Minotti, P.L. (1988). Increasing yield of tomatoes with plastic mulch and apex removal. *Journal of the American Society of Horticultural Science* 113: 342-347.

Wellso, S.G. and Araya, J.E. (1993). Resistance stability of the secondary tiller of 'Caldwell' wheat after the primary culm was infested with virulent Hessian fly (Diptera: Cecidomyiidae) larvae. *The Great Lakes Entomologist* 26: 71-76.

Wellso, S.G. and Hoxie, R.P. (1994). Tillering response of 'Monon' and 'Newton' winter wheats infested with biotype L Hessian fly (Diptera: Cecidomyiidae) larvae. *The Great Lakes Entomologist* 27: 235-239.

Chapter 9

Breeding Maize
to Face Weed Problems

Claudio M. Ghersa

INTRODUCTION

In modern agriculture, weed problems have been solved by using technologies that rely on soil tillage and, increasingly, on herbicide spraying (Radosevich, Holt, and Ghersa, 1997). These technologies proved to be very efficient; therefore, there was no need to include in the breeding programs selection for better crop performance under weed competition. This situation has changed in recent years, with increasing social pressure on farmers to reduce soil erosion and use of agrochemicals, and an alarming occurrence of herbicide-resistant weeds. These factors have pushed researchers to start breeding programs aimed to improve the ability of crops to produce in competition with weeds (Kropff and Lotz, 1992; Callaway and Forcella, 1993; Powles and Holtum, 1994).

Also in the last decade, availability of new tools for genetic manipulation has allowed breeders to develop "ideological" selection programs to complement the traditional ones based on empirical selection; ideological selection means that plant breeders improve crop yield by design rather than by empirical selection (Evans, 1993). Examples of ideological selection are the recently available maize crops resistant to herbicides and insect attack (Hunter, 1993).

Considering these facts, the identification of some of the main processes that are involved in maize production, and understanding how weed interference may affect them, can help design new maize crops. The new ideotype can make use of the advantage of having traits that

157

reduce the negative effects of weed interference on yield potential. The objectives of this chapter are to identify those traits valuable for facing weed problems in maize production and to discuss their usefulness in particular management and ecological scenarios.

WEED INTERFERENCE

Weed interference in maize crop production can be simplified and divided into two periods: (1) presowing/preemergence and (2) during the crop growing cycle. During the first period, the processes of depletion of soil resources and/or soil contamination by weeds are considered. Water and nitrogen are, frequently, the most important resources depleted by growing plants, and liberation of allelochemicals to the soil is important for contamination. However, although rarely mentioned in the literature, accumulation of live or dead biomass should also be considered (Radosevich, Holt, and Ghersa, 1997).

In the second period, when weeds and the crop are actively growing, competition for light, soil water, and nutrients is the most important process controlling crop yield (Spitters and Aerts, 1983), but, in some cases, allelopathy can appear as an important factor curtailing crop yield (Radosevich, Holt, and Ghersa, 1997).

MANAGEMENT SCENARIOS AND STRATEGIES

Management strategies are designed to reduce the negative effects of these processes on yield and are sustained in two different paradigms: the first one, and most popular in modern agriculture, considers management as the implementation of technologies to control the crop environment to achieve optimal production. This type of strategy requires skilled managers and hybrids or varieties with "rigid phenotype" that vary little when grown in different environments. This means, for example, that maize genotypes will express a relatively low-density-dependent response and have a stable harvest index (Tollenaar, Dwyer, and Stewart, 1992; Andrade et al., 1996). Crop optimal performance will depend on adequate seeding procedure (soil seedbed preparation, timing, seed density, and spatial arrangement), soil fertilization, irrigation, and so forth.

The second paradigm considers a scenario with a less skillful or less resourceful manager, thus assuming that the crop environment is poorly controlled and that optimum production will depend more upon the capability of the crop plants to adjust to it. A changing, "plastic phenotype" is required for these conditions.

CROP DESIGN

Rigid Phenotype

Interference During the Presowing/Preemergence Period

When weeds are allowed to grow before the establishment of a crop, they will use soil water and nutrients and accumulate live and dead biomass as well as liberate allelotoxins. Obviously, the best management decision would be to control weeds before these processes advance to a stage during which they will reduce maize production significantly. Nevertheless, it is quite common to find situations in which weeds grow before planting. In rain-fed production systems and with low levels of fertilizer use, such as in many farms of Argentina's west Pampa region, weeds growing before the crop is planted appear to be a problem. If this problem relates only to soil water or nitrogen depletion, then it can be viewed as scarce resource availability due to soil and climate constraints. Therefore, traits that help reduce soil evaporation and improve crop water use efficiency (e.g., osmotic adjustment) will improve yields (Turner, 1993). On the other hand, if the problem relates to toxin accumulations in the soil, evidence shows variability in the response to toxins for many organisms, including plants (Walton et al., 1994). Taking this into account, there is no reason why genes conferring specific toxin tolerance or soil detoxification by the plant rhizosphere could not be incorporated into maize plants.

The interference problem can also be related to accumulation of weed dead biomass on the soil surface. If sowing is possible under this condition, a frequent problem derived from soil cover by plant residues is the reduction in soil surface temperature. Low temperatures in the soil surface layer curtail the rate of germination and emergence of the crop seedlings, making them more vulnerable to attack by insects and diseases, as well as reducing the ability of the crop to compete against weeds (Spitters, 1989). In recent years, related to the expansion of no tillage and to the extension of the cropping season, a great breeding effort has attempted to provide farmers with maize hybrids able to germinate and emerge adequately under low soil temperatures (Giauffret, Bonhomme, and Derieux, 1995). These efforts have been fruitful, but, probably, some room remains for improvement in this field.

Interference During the Crop Growing Season

Timing between crop and weed emergence is a key factor controlling competition (Kropff and Spitters, 1991). Precise forecasts of weed emergence from the soil surface, for both annuals and perennials, can be carried out using thermal time calendars (Ghersa and Holt, 1995; Satorre, Rizzo, and Arias, 1996), allowing for selection of a maize crop and planting date that will minimize the negative effects of weed interference on maize production (Ghersa and Martinez-Ghersa, 1991).

Growth rate, leaf expansion, plant height, and branching contribute to the architecture of the plant canopy and, consequently, to how quickly and how effectively the crop shades the weeds. Maize, because of its reduced branching, the erectofile characteristic of its leaves, and its plant height, presents a low light extinction rate and frequently allows light to reach the soil surface. Recent maize breeding programs have reduced plant height and modified the angle of the leaves (Maddonni and Otegui, 1996). These traits are desirable to reduce weed competition for light. For example, bermudagrass growth is highly affected when the light reaching the soil surface is reduced beyond 65 percent (Solari et al., 1997).

Root growth, root distribution in the soil profile, and root system functionality have received considerably less attention from scientists than the aboveground parts. Nevertheless, enough information exists to consider that dense and deep rooting systems provide plants with a better competitive ability, this trait being particularly important for rain-fed systems. Cárcova (1998) has studied the rooting system of maize hybrids and has found great differences among them. It can be speculated that the one which will suffer less stress under drought conditions is the one having both a rooting pattern that is relatively less affected by the soil physical characteristics and a higher root density. In spite of the fact that these traits have not been evaluated for their value in relation to competitive ability against weeds, presumably, a maize hybrid with these characteristics will perform better under weed competition for soil resources.

Plastic Phenotype

Designing the plastic ideotype appears to be intellectually challenging and opens the possibility for considering whether an "all ground" crop can really be designed, and if such a creation would be useful for farming. My prejudgment is that "all-use tools" are frequently useless marketing artifacts, and that they are hardly ever used in a workshop. Therefore, the ideas discussed under this topic are presented considering the analogy with a variable size wrench rather than an all-use tool. This means that traits conferring phenotypic plasticity should be examined within the context of a particular function, such as adjusting nuts of various sizes, the self-regulating size of the wrench's mouth being the plastic trait to be engineered.

Interference During the Presowing/Preemergence Period

An interesting aspect of phenotypic plasticity found in plant seeds is related to germination response. Seeds can be dormant and released from dormancy with particular signals, such as a particular range of low temperatures or of light quality/intensity. Whether incorporating dormancy traits in maize can be useful in avoiding weed interference has yet to be tested. Nevertheless, it can be argued that, in some ecological scenarios, in which relay cropping is practiced, seed dormancy could have a role (Ghersa, Martinez-Ghersa, and Benech-Arnold 1997). The cropping system would consist of sowing maize as a second crop together with the first one. Early germination of maize, while the first crop is still growing, could allow continuous leaf cover and root occupation of the soil profile. This cropping system would not reduce soil depletion of water and nutrients for maize, but it would stop their use by weeds. Also, by selecting the adequate first crop, the amount of water and nutrients used by it could be kept somewhat under control.

Interference During the Crop Growing Season

The ideal plasticity for a self-designing canopy architecture would be the one capable of capturing all the photosynthetically active radiation (PAR) and, therefore, able to respond to the presence of a gap by covering it with green tissue at low energy costs. This would benefit the crop by capturing more energy and would provide an inadequate light environment for weed growth. Weeds are able to respond in this way, and some species are even able to elongate their internodes and concentrate their leaves, always overtopping the ones of the crop (Radosevich, Holt, and Ghersa, 1997).

Callaway and Forcella (1993) selected soybean varieties considering high leaf area expansion rate during the seedling stage of growth, and they found that these selections could tolerate higher weed densities without losing yield. Maize hybrids with this trait could respond in a similar way.

Incorporating traits for plasticity in stem elongation into crops would be inadequate. One of the major achievements in plant breeding was to avoid lodging by eliminating stem elongation response to increasing density. Taking this into account, it is possible to think that a crop able to orient its leaves toward canopy gaps, without major changes in plant height, could increase the asymmetry in crop competition with late-emerging weeds and with creeping weeds (Solari et al., 1997).

Branching is highly responsive to light quality and intensity, and it varies with plant overcrowding. Is it, therefore, an important trait to consider in maize improvement? The answer to this question has already been addressed by many researchers. Undoubtably, failures in seeding or irregularities in seeding rate are frequently important factors determining opening of gaps in the crop canopy, and these gaps are readily used by the weeds to grow. Tiller formation traits are helpful in solving this problem, as long as they do not reduce the harvest index or generate a heterogeneous kernel maturation.

Potential for osmotic adjustment may be another trait to evaluate, to determine whether it can improve competitive ability of maize plants. Lemcoff, Chimenti, and Davezac (1998) have studied this trait in maize hybrids and found that there is some benefit for those capable of osmotic adjustment when plants were stressed by drought.

More recent experiments show that *Eucalyptus* sp. seedlings selected for their capability of osmotic adjustment performed better than those lacking this characteristic when competing with herbaceous weeds (A. Garau, personal communication), and it is therefore possible to think that maize could behave in a similar way.

Some researchers have considered perennial maize as an ideotype (Jackson 1990), and it is probably in a perennial breed where many traits conferring plasticity could help in generating a monospecific canopy, or at least one highly dominated by the crop. The aim in a breeding program should be to achieve maize plants with an opportunistic strategy. This would mean that the vegetative part of the plant should support stress without dying, as well as being able to efficiently capture resources and produce heavy seed yield when resources become available. Such plant behavior is found in some of maize weedy relatives (*Theosinte* spp.) and, therefore, it could be possible to transfer some of their traits into maize. In a perennial plant type, traits conferring plasticity for root distribution and liberation of root exudations could also be incorporated. Root exudations have been described for various dominant species, such as *Helianthus annuus* and *Sorghum halepense,* and they have been considered as having a major role in generating and sustaining the species' dominance in the community (Rice, 1984).

CONCLUDING REMARKS

Weeds are still a problem in maize production systems, and some room is left for plant breeders to improve maize crop characteristics aimed at reducing the negative effects of weed interference on yield.

Improvement based on ideological models, which in turn can be tested using mathematical modeling, can reduce the costs of empirical selection and yield better results.

The agronomic/technological/ecological systems have to be adequately described to identify which are the desirable traits to incorporate to achieve a realistic improvement.

REFERENCES

Andrade, F.H., Cirilo, A.G., Uhart, S.A., and Otegui, M.E. (1996). *Ecofisiología del cultivo de maíz.* Balcarce, Argentina: La Barrosa, CERBAS, and Dekalb Press.

Callaway, C.B. and Forcella, F. (1993). Crop tolerance to weeds. In Callaway, M.B. and Francis, C. (Eds.), *Crop improvement for sustainable agriculture* (pp. 100-131). Lincoln, NE: University of Nebraska Press.

Cárcova, J. (1998). Comportamiento del sistema radical de tres genotipos de maíz en ambientes de la Pampa ondulada con distinto deterioro. MSc thesis, Universidad de Buenos Aires, Buenos Aires, Argentina.

Evans, L.T. (1993). Processes, genes, and yield potential. In Buxton, D.R., Shibles, R., Forsberg, R.A., Blad, B.L., Asay, K.H., Paulsen, G.M., and Wilson, R.F. (Eds.), *International crop science,* Volume I (pp. 687-696). Madison, WI: Crop Science Society of America.

Ghersa, C.M. and Holt, J.S. (1995). Using phenology prediction in weed management: A review. *Weed Research* 35: 461-470.

Ghersa, C.M. and Martinez-Ghersa, M.A. (1991). A field method for predicting yield losses in maize caused by Johnsongrass (*Sorghum halepense*). *Weed Technology* 5: 279-285.

Ghersa, C.M., Martinez-Ghersa, M.A., and Benech-Arnold, R.L. (1997). Seed dormancy implications in grain and forage production. *Journal of Production Agriculture* 10: 111-117.

Giauffret, C., Bonhomme, R., and Derieux, M. (1995). Genotypic differences for temperature response of leaf appearence rate and leaf elongation rate in field-grown maize. *Agronomie* 15: 123-137.

Hunter, R.B. (1993). The molecular biologist and the plant breeder. In Buxton, D.R., Shibles, R., Forsberg, R.A., Blad, B.L., Asay, K.H., Paulsen, G.M., and Wilson, R.F. (Eds.), *International crop science,* Volume I (pp. 529-531). Madison, WI: Crop Science Society of America.

Jackson, W. (1990). Searching for high seed yielding herbaceous perennials. In Francis, C.A., Butler Flora, C., and King, L.D. (Eds.), *Sustainable agriculture in temperate zones* (pp. 384-393). New York: John Wiley and Sons, Inc.

Kropff, M.J. and Lotz, L.A.P. (1992). Systems approaches to quantify crop-weed interactions and their application in weed management. *Agricultural Systems* 40: 265-282.

Kropff, M.J. and Spitters, C.J.T. (1991). A simple model of crop loss by weed competition from early observations on relative leaf area of the weeds. *Weed Research* 31: 97-105.

Lemcoff, J.H., Chimenti, C.A., and Davezac, T.A.E. (1998). Osmotic adjustment in maize (*Zea mays* L.): Changes with ontogeny and its relationship with phenotypic stability. *Journal of Agronomy and Crop Science* 180: 241-247.

Maddonni, G.A. and Otegui, M.E. (1996). Leaf area, light interception, and crop development in maize. *Field Crops Research* 48: 81-87.

Powles, S.B. and Holtum, J.A.M. (1994). *Herbicide resistance in plants: Biology and biochemistry.* Boca Raton, FL: Lewis Publishers.

Radosevich, S.R., Holt, J.S., and Ghersa, C.M. (1997). *Weed ecology: Implications for management.* New York: John Wiley and Sons, Inc.

Rice, E.L. (1984). *Allelopathy.* Orlando, FL: Academic Press.

Satorre, E.H., Rizzo, F.A., and Arias, S.P. (1996). The effect of temperature on sprouting and early establishment of *Cynodon dactylon. Weed Research* 36: 431-440.

Solari, F., Satorre, E.H., Arias, S.P., and Guglielmini, A. (1997). Efecto del canopeo de soja sobre el comportamiento del gramón (*Cynodon dactylon* L. Pers.). *Actas XVIII reunion Argentina de ecologia* (p. 119). September 17-19. Buenos Aires, Argentina: Asociación Argentina de Ecología.

Spitters, C.J.T. (1989). Weeds: Population dynamics, germination and competition. In Rabbinge, R., Ward, S.A., and van Laar, H.H. (Eds.), *Simulation and systems management in crop protection* (pp. 182-216). Simulation Monographs #32. Wageningen, Netherlands: Pudoc.

Spitters, C.J.T. and Aerts, R. (1983). Simulation of competition for light and water in crop-weed associations. *Aspects of Applied Biology* 4: 467-483.

Tollenaar, M., Dwyer, L.M., and Stewart, D.W. (1992). Ear and kernel formation in maize hybrids representing three decades of grain yield improvement in Ontario. *Crop Science* 32: 432-438.

Turner, N.C. (1993). Water use efficiency of crop plants: Potential for improvement. In Buxton, D.R., Shibles, R., Forsberg, R.A., Blad, B.L., Asay, K.H., Paulsen, G.M., and Wilson, R.F. (Eds.), *International crop science,* Volume I (pp. 75-82). Madison, WI: Crop Science Society of America.

Walton, B.T., Hoylman, A.M., Perez, M.M., Anderson, T.A., Johnson, T.R., Guthrie, E.A., and Christman, R.F. (1994). Rhizophere microbial communities as a plant defense against toxic substances in soils. *Proceedings of the ACS, Symposium Series* 7: 82-92.

Chapter 10

Use of Simulation Models for Crop Improvement

Anthony Hunt

INTRODUCTION

Improvement in the yield potential and quality of many crop species has been striking over the past few decades, as documented almost ad nauseam in many publications. Such improvement was achieved in virtually all cases by sustained direct selection rather than by indirect selection for processes that are important in the yield elaboration process. Improvement in yield stability/reliability also appears to have been impressive in some crops, although this aspect has not been as well documented. In this case, direct selection has been complemented by indirect selection for aspects that contribute to yield stability, such as pest and disease resistance. These latter are readily observable characteristics, as were many of those which played a role in the domestication process. They thus contrast with the physiological characteristics that underlie yield potential, which are not readily observable (Frankel, 1947), perhaps not even known, and certainly not agreed upon.

The physiological characteristics that determine yield potential, as Evans (1993) has pointed out, may never be universally agreed upon because crop yield is the ultimate outcome of the whole life cycle of the crop, and of the rates, durations, and interlinkages of many processes at all stages of development. No single process provides the master key to greater yield potential under all environmental conditions. Quite different processes may limit the yield of different cultivars at one site and in one season, or of one cultivar at different sites or in different seasons (Crosbie and Mock, 1981; Willman et al., 1987). Yield potential is an extremely complex characteristic, the improvement of which, in most cases, demands an integrated change in a number of physiological processes, and certainly a balanced change in processes related to both source and sink aspects, it being of little value to increase the sink if the source is not concurrently augmented, and vice versa.

In contrast to considerations of the characteristics per se is the hope that consideration of the framework in which the various characteristics operate can be agreed upon, and that the interrelationships and interdependencies of characteristics can be more clearly specified. Such a framework, which could be developed in a purely schematic, qualitative manner (e.g., Warren-Wilson, 1972), will achieve its greatest utility when developed such that the impact of different interacting scenarios can be examined quantitatively. In such a form, the framework becomes a mechanistic crop simulation model. Although a number of these already exist, few contain sufficient detail to allow them to be rigorously applied to examine different hypotheses concerning the operation of the whole crop system, in which interplant competition is always significant. Several, however, have been applied to examine the value of one characteristic in one environment and thus extend the "master key" type of thinking that has underlain much past work aimed at unraveling the physiological basis of yield.

The availability of mechanistic simulation models that account for some process interactions, however, ensures that at least some of the current physiological knowledge is encapsulated in a manner that makes it useful to those intimately concerned with crop improvement, those who must focus on the plant as an integrated entity. Given this situation, it seems timely to examine the use (potential?) of simulation models in crop improvement, and to build on previous efforts in the domain (e.g., Aggarwal et al., 1997; Hammer et al., 1997; Hunt, 1993; Muchow and Kropff, 1997). In so doing, however, it will be important to remember that plant breeders have been subjected over the past few years to a barrage of suggestions from nonbreeders as to how to improve their activities, suggestions that, in some cases, have taken little account of the realities of the plant breeding process. As a starting point, therefore, some aspects of the overall crop improvement process will be highlighted.

THE CROP IMPROVEMENT PROCESS

The goal of a crop improvement program is the development of new, improved cultivars or breeding lines/populations for particular target areas and for specific applications. The breeding process is thus always concerned with the creation of something new, with building on what currently exists to create something better. Attainment of the goal often represents a time investment of from ten to fifteen or more years. Because of this, breeders are continually and acutely aware not only of the need for constant evaluation and refinement of procedures but also of the cost of following procedures that may not be particularly efficient in the production of superior cultivars. Breeders are also aware that superior cultivars can be developed using different breeding pathways, that there may not be one "best" method or even outcome, and that they will generally have to be satisfied with an improvement of a few percent. Because of this latter, they have to use, at least in the final breeding phase, techniques that can detect small differences. Breeders are also acutely aware of time constraints, particularly in the period between harvesting and seeding, and of the need for data handling and analytical techniques that can not only be applied to large numbers but that also work quickly and efficiently in a routine setting.

The situations and options that can lead to an efficient breeding process thus present fascinating challenges to the breeder. Many of these challenges resolve themselves into questions of resource distribution among different aspects of the program. For a self-pollinated cereal, these questions are best considered by dividing the breeding process into three phases, each of which involves the breeder in quite different functions. The entire process can be divided first into heterozygous and homozygous phases, with the heterozygous phase dealing with everything up to individual line selection and the homozygous phase dealing with all subsequent line evaluations and cultivar release. The heterozygous phase can be further subdivided into a planning and hybridization phase, and one concerned with the handling of the early generation progenies. The personal time of the breeder may be divided among these three phases in different ways—in one case, with a wheat breeding program, Jensen (1975) estimated that the breeder's time was distributed in a 40-10-50 percent fashion among the phases, while the division of the time of the technical "crew" was quite different, on the order of 5-10-85 percent (see Table 10.1).

For a cross-pollinated crop, the process is somewhat different, and an element of recurrency is often introduced before material is suitable for release as a cultivar (see Figure 10.1). The distribution of the breeder's time, however, may be quite similar to that for a self-pollinated crop, while the distribution of the technical crew's time would be increased for the hybridization and segregation phases. For cross-pollinated crops in which hybrid systems are available and used, the process has further complications, insofar as heterotic patterns must be considered and evaluated.

Although the actual figures on distribution of effort vary among programs, all programs show a large percent of the breeder's time assigned to the planning phase. Such a distribution reflects the fact that most breeders consider this phase of critical importance to the creation of a superior genotype. Indeed, most breeders have traditionally spent time on the design of the desired genotype for their target area, and on the siting of some of their testing locations, although rarely on the siting of their main breeding locations, which

TABLE 10.1. Division of Time Among the Different Phases of Activity in a Self-Pollinated Crop Breeding Program

Phase	Breeder's time	Technicians' time
	Percent	
1. Planning and hybrid-ization (heterozygous)	40	5
2. Segregation and sta-bilization	10	10
3. Line evaluation and release (homozygous)	50	85
	100	100

Source: Adapted from Jensen (1975).

FIGURE 10.1. Phases Involved in Breeding a Cross-Pollinated Crop

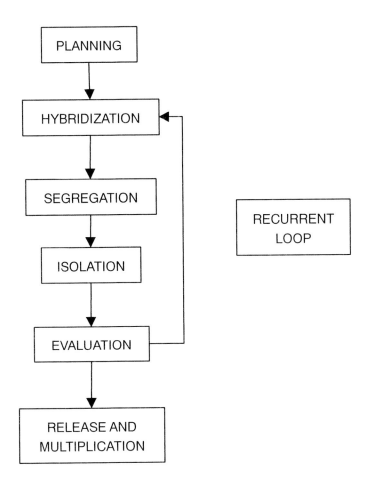

are generally fixed by historical and "political" considerations. The design planning has not only involved consideration of the functional requirements, both for yield potential and for the specific requirements of the target environment (e.g., appropriate maturity, lodging resistance, disease resistance, tolerance of temperature extremes or mineral deficiency or excess, kernel size, grain yield and quality, absence of annoying "rough" awns), but has also involved consideration of parents that could be used as a source of specific traits. Consideration must be made also of the most appropriate methods for using parents that may only have one desirable attribute as well as those which are generally good performers in the target area, and of testing procedures that could be used to identify desirable segregants. In these considerations, the focus has always been on current and future limitations to performance, and on ways to overcome them, so that information which is relevant in this context is of more value than historical information indicating what has been achieved in the past. This aspect is perhaps of even greater significance in the current era of change in climate, in management systems (e.g., herbicides, seed treatments, tillage, fertilizer), and in new market opportunities (e.g., high oil corn, low phytate corn).

Even with careful choice of parental materials, the fact remains that the chances of success in a breeding program depend also on the number of lines evaluated each year. With 50 percent of the breeder's time and 85 percent of the technical crew's time devoted to this aspect, it is obvious that techniques which could help reduce resource demands during the evaluation phase would be of tremendous value. The large demands during this phase stem from several different problems. First, it is generally impossible to decide which breeding lines are superior on the basis of data obtained with plants growing individually (spaced plants), so evaluation of their true potential not only has to wait until there is sufficient seed available for a plot scale test but also has to be undertaken in a fashion that demands a lot of labor and general attention to crop management. Second, data obtained in a trial at one site are usually not indicative of what would happen in a trial at another site in the same year, or even at the same site in another year. Because of these problems, usually referred to as genotype × location and genotype × year interactions (or, in general, genotype × environment interactions), the evaluation process generally has to be continued at several sites and for several years. Third, the material that is produced as a result of the plant breeding process generally does not match up in all aspects to the breeding design ideal. Because of this, the breeder has to spend further time deciding which of the available breeding lines, each one a different combination of required traits, is likely to perform best for the seasons that will follow the testing process. A classical example of the conundrum arises when one line proves to be high yielding but of low disease resistance, and another line to be lower yielding but of higher disease resistance. In such a case, the breeder must determine whether the disease resistance requirements in the years following testing, and on the farm rather than in a research plot, are likely to be greater or less than during testing. In essence, the breeder must decide whether the line that was lower yielding during testing might, because of its disease resistance, be higher yielding on the farm and in the years subsequent to testing.

PLANNING: YIELD POTENTIAL

Although a consideration of the characteristics needed for high yield has been a part of the parental selection process from the time of its inception as an activity, Donald (1962), in a classic article titled "In Search of Yield," argued that too little attention had been paid to the basic processes governing dry-matter production and its transformation into economic yield, and to characteristics governing these processes. Donald (1968a, b), expanded these arguments and developed a breeding approach based much more explicitly on the design of model plants or ideotypes than on traditional breeding. The basis of this approach was the use of known principles of physiology and agronomy to design a plant that was capable of greater production than existing types. Donald also suggested that such a model plant was likely to involve a combination of characteristics that would rarely, if ever, occur by chance in breeders' plots. Characteristics highlighted were largely morphological: a short stem, a few small erect leaves, a high harvest index, an erect ear or panicle, awns, and a single culm. Donald (1979) later reported an attempt to develop such an ideotype in barley. The lines developed were uniculm, six-rowed, awned barleys, with a high harvest index relative to the commercial cultivars available at that time. However, they had lax medium-length leaves rather than the short erect leaves.

The work of Donald (loc cit) has been followed by several attempts to define the characteristics needed for high and stable yield (see Austin, 1975; Lawn and Imrie, 1991; Marshall, 1991; Muchow, Hammer, and Carberry, 1991; Shorter, Lawn, and Hammer, 1991; Hunt, 1993), attempts that have most recently involved computer simulation approaches and nonmorphological traits. Examples using simulation models and dealing with traits such as rooting depth, crop height, maximum leaf area, harvest index, maturity class, temperature response, water stress sensitivity at different stages, osmoregulation, and stomatal sensitivity to water stress were published for a number of environments and crops in the 1970s and 1980s (e.g., Field and Hunt, 1974; Jones and Zur, 1982; Jordan, Dugas, and Shouse, 1983; ICARDA, 1986). In one such study, O'Toole and Jones (1987) reported an application of the simulation model AL-MANAC, which is a submodel of EPIC (the Erosion-Productivity Impact Calculator; Sharpley and Williams, 1990), to evaluate traits needed for high and consistent productivity in rain-fed rice crops in parts of India. The traits considered and the values examined are shown in Table 10.2, and twenty-year simulation results are in Table 10.3. These showed that mean yields for 140-day varieties were low and highly variable, whereas a 105-day intermediate type maintained a mean yield close to that of high-yielding varieties, coupled with a standard deviation near that of the stable traditional materials.

Subsequently, O'Toole and Stockle (1991) presented two examples of the use of simulation models for analyzing temperature-related traits: root growth and grain filling. In this work, the spring wheat model developed by Stockle (1985) was used, and growth and yield of dryland spring wheat grown in eastern Washington (United States), where the model was calibrated, was simulated. The authors ran simulations with optimal root temperatures varying from 18°C to 26°C. Root depths, as a function of time for hypothetical cultivars that perform optimally in soils at 20 and 26°C were computed. Results showed that the cultivar with 20°C as the optimum soil temperature had an advantage because its roots reached a given depth sooner and because they explored the soil profile deeply. The impact of the temperature range considered for root growth on grain yield was also clear (see Figure 10.2). Yield (and rooting depth) was maximum in soils at 18°C, but decreased slightly at 20°C, and sharply at higher temperatures. Terminal water stress was the cause of yield reduction for cultivars with roots that performed optimally at higher temperatures. The soil was too cold for roots to penetrate deep enough to extract the water available. The degree of water stress during grain filling increased accordingly.

TABLE 10.2. Genotype Traits Considered by O'Toole and Jones (1987) in a Simulation of Rain-Fed Rice Production in India

Development	
Total life cycle duration	90-140 days; 1850-2850 °C days >10°C
Period to 50 percent maximum leaf area index	0.55-0.71 of total
Period to postanthesis leaf senescence	0.67-0.71 of total
Period to onset of critical water stress phase	0.55-0.71 of total
Period to end of critical water stress phase	0.78-0.86 of total
Growth	
Standard harvest index	0.29-0.42
Maximum leaf area index	4.0-7.5 $m^2 \cdot m^{-2}$
Maximum crop height	0.75-1.50 m
Maximum root depth	0.7-1.5 m
Water stress effect on radiation conversion	1.5-2.0
Water stress effect on harvest index	1.5-2.0
Minimum harvest index	0.01

Source: O'Toole and Jones (1987).

TABLE 10.3. Average Yields from Twenty-Year Simulations of Nine Combinations of Plant Type and Maturity Group

Plant type	Average yields (tons/ha[1] ± standard deviation)		
	90 days	**105 days**	**140 days**
High yield	1.98 + 0.60	2.02 + 0.99	1.32 + 1.15
Intermediate	1.82 + 0.22	2.19 + 0.56	1.50 + 1.21
Traditional	1.26 + 0.08	1.69 + 0.38	1.22 + 0.98

Source: O'Toole and Jones (1987).

FIGURE 10.2. Simulated Root Penetration (Triangles) at the End of the Season and Grain Yield (Circles) with Different Root Growth Optimum Temperatures

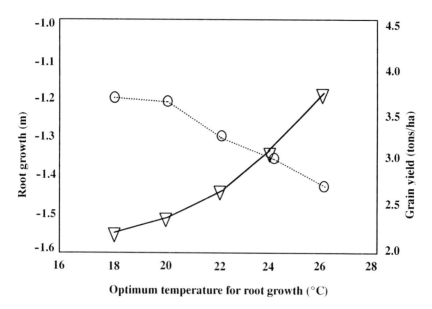

Optimum temperature for root growth (°C)

Source: O'Toole and Stockle (1991).

More recently, simulation models have been used in attempts to identify traits required in a high-yielding rice of the future (Penning de Vries, 1991; Dingkuhn et al., 1991), one that may be grown under different agronomic conditions (direct seeded) than in the past (transplanted). Traits identified as desirable have included the following: (1) enhanced leaf growth during crop establishment, (2) reduced tillering, (3) less foliar growth and enhanced assimilate export to stems during late vegetative and reproductive growth, (4) sustained high foliar nitrogen (N) concentration, (5) a steeper slope of N concentration from the upper to the lower leaf canopy layers, (6) expanded capacity of stems to store assimilates, and (7) a prolonged grain-filling period (Dingkuhn et al., 1991). Such a list, in which some aspects are not easily translated into plant characteristics with which a plant breeder can work, is reminiscent of the original ideotype lists of Donald and others. Its use in practical plant breeding would suffer from the same limitations identified for the ideotype concept from more than twenty years of effort. Nonetheless, the use of simulation models in this context provides a mechanism for quantitatively evaluating the impact of hypothetical changes, and it is one that has continued to be demonstrated and extended (e.g., ten Berge, Aggarwal, and Kropff, 1997), with some attention even being paid to aspects related to weed competition (Dingkuhn et al., 1997; Lindquist and Kropff, 1997).

PLANNING: YIELD GAPS

Although knowledge of yield potential traits is of more than passing interest to those directly concerned with crop improvement, knowledge of environments, environmental aspects, and of the associated plant responses that cause performance to be less than potential at some locations within a target breeding area is generally of much more interest. Identification of these environments, and of the reasons for such poor performance, is of critical importance to the success of a crop improvement endeavor. Simulation modeling can play a continuing key role in this domain by providing information on "theoretical" yields attainable at specific trial sites, given the seasonal radiation, temperature, and rainfall conditions; the management applied to the crop; and the soil characteristics. A comparison of actual and theoretical yields—a "yield gap" analysis—can then help indicate whether the yields achieved by both the best-performing cultivar/breeding line at a trial site and the other cultivars/lines reflect the impact of the seasonal weather conditions (but within the context of those incorporated into the model) or non-weather-related causes.

An example of such a potential application is provided by studies conducted at eight or more sites in India, all using one standard peanut cultivar (Boote, Jones, and Singh, 1991; Singh et al., 1994). The model-predicted climatic yield potential was achieved at about one-third of these sites, but at many locations, poor growth and low yields could not be attributed to weather conditions. It was concluded that other factors, such as soil fertility and pests, were causing a yield gap at some sites, and this led to increased research emphasis on these problems.

A second example in which model outputs were used to provide surrogate "check" variety yields against which actual yields in a multilocation breeding trial could be compared was provided by Saulescu and Kronstad (1995) for wheat in the Pacific Northwest of the United States. These authors presented the trial data in terms of differences between the actual yield of a cultivar or breeding line and the yield either of the surrogate check or of a real check (Stephens), and in terms of interaction variances between the checks and the trial entries. When the simulated yield was used as a check, interactions for some cultivars were greater than when using real data (see Figure 10.3), possibly reflecting the fact that some of the important causes

FIGURE 10.3. Interaction Variances for Yield [(t/ha)^2] Between Check Cultivars and Trial Entries in Multilocation Wheat Trials in Oregon, United States

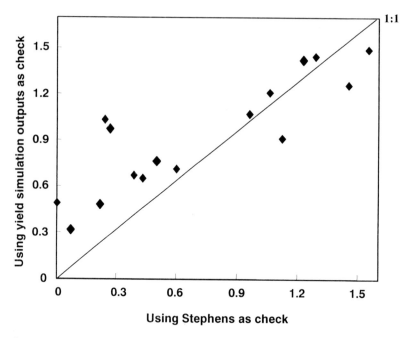

Source: Adapted from Saulescu and Kronstad (1995, pp. 775-776).

Note: Check yields were either simulated using the MODWHT3 model or taken from data for the variety Stephens.

of genotype × environment interaction (e.g., diseases, lodging) were not taken into account in the simulation model. All of the interaction variances using simulated data were significant, however, suggesting that the majority of the genotypes had a reaction to environment that was different from that embedded in the model. Further, many of the model-derived variances fell close to the 1:1 line when plotted against those variances obtained using real data, as would be expected given that the model used (MODWHT3; Rickman, Waldman, and Klepper, 1996) was calibrated for the genetic characteristics of the real check variety Stephens.

The previous considerations have been concerned with the use of models as they come from the originator, with the user merely generating the appropriate input files (soil, weather, crop management, and cultivar-specific characteristics). A second major domain in which models can be used to help in the interpretation of yield gap data, and one that has not yet been widely adopted, is through the provision of a "test bed" into which new hypotheses dealing with the reasons for genotype × environment interactions can be incorporated and evaluated by those who were responsible for generating the data and making the interpretations. Such an approach has been followed, for example, by Wagner-Riddle and colleagues (1997), who modified the water balance subroutine of the Soygro model (Jones et al., 1989) to account for surface mulch when endeavoring to interpret data from a cropping system study.

It has been argued (e.g., Sinclair and Seligman, 1996) that benefit in this latter domain will probably be greater when the modeling approach is not prejudiced by automatically using existing models. However, the problems of accommodating the intricacies of input-output, and of achieving a balanced consideration of all aspects that impact significantly on performance in most cropping situations, are generally likely to render a de novo approach impractical. The alternative, the use of an existing model, however, will only be possible when users not only have an understanding of the details of the model but are also able to make changes to the model with relative ease. Steps currently being taken to emphasize "modularity" and simplification (e.g., by International Consortium for Agricultural Systems Applications [ICASA]), and to restructure some existing models (e.g., the GRO series of models; J. W. Jones, personal communication), will help in this regard. Further efforts along these lines, as well as efforts by breeders (and physiologists!) to learn the details of some of the existing models, will be necessary before this aid to interpretation of yield gap information can be widely used.

One cautionary note is necessary here, however, because simulation results depend on the accuracy of the input data, in particular, on the weather information. Detection of a "yield gap," or of differences in relative performance from site to site, should thus not only stimulate further thought on agronomic aspects but also on environmental data. Boote, Jones, and Pickering (1996), for example, reported on peanut studies at two sites about 20 kilometers (km) apart near Gainesville, Florida. Weather data were collected at each site, and the PNUTGRO model (Boote et al., 1989) was used to predict growth at both locations. The model predicted a lower biomass increase at one site, even though measured growth was similar at both sites. Comparison of the solar radiation data indicated that one sensor was providing values 15 percent lower than those recorded at the other site. The sensor at this site had apparently deteriorated with time. Careful checking of weather-recording instruments, particularly those dealing with solar radiation, is thus essential for application of yield gap analysis to multilocation trials.

LINE EVALUATION

A major problem in crop improvement lies in the fact that final yields measured at one location for a new breeding line or cultivar often provide little insight into how the genotype will perform relative to other genotypes at different locations or in different years. How to overcome this problem of genotype × environment interaction when endeavoring to draw conclusions from multilocation trials has been the subject of much effort (see Cooper and Hammer, 1997; Yan and Hunt, 1998). One suggestion has been to weigh the results from individual trials in accordance with the frequency with which the environment at the intended trial site figures in the overall array of environments over years and sites in the target region (Muchow, Cooper, and Hammer, 1996). The implementation of such a suggestion requires good environmental characterization at many sites in a target region, and good establishment of frequency distribution over time and space for important environmental characteristics. Primary environmental statistics could be used, but because of the interacting effects of rainfall and temperatures, soil type, and initial soil water contents, it may be better to use derived environmental characteristics that take these aspects into account. Muchow, Cooper, and Hammer (1996) followed this line of reasoning and explored the use of simulation models to provide water-related indices of relevance to crop performance in a dryland environment. Their analysis showed that indices of water deficit generated by a crop growth simulation model could be used to show how the water regime in a specific experiment is related to the array of regimes encountered over space and time in a target area. Calculation of weekly indices of water deficit based on climatic data, and classifying the type of season on the basis of these indices, successfully identified groups having distinct patterns of development of water deficit. The frequency of occurrence of the different groups varied, thus allowing the environmental conditions experienced in a given season to be weighted in terms of long-term environmental conditions at that location.

A second approach would be to use a simulation model directly to extend experimental results from one or a few sites and years to many, and even to compare the outcomes from different management strategies when applied to different genotypes. This is a major potential application that allows for the use of the historic weather and soil data of a region, and that facilitates a consideration of the probability aspects associated with particular genotypes and management regimes (Singh and Thornton, 1992). Its use, however, rests on the precision and/or accuracy of the model being used. Models are able to explain a significant proportion of the variation in performance of standard well-adapted cultivars under different environmental conditions (e.g., Muchow and Bellamy, 1991). They have been used successfully to assess production risk in some environments (e.g., Carberry, Muchow, and McCown, 1993) and to aid crop management decisions, again in some specific environments (e.g., N application in variable rainfall environments; Keating, Godwin, and Watiki, 1991). Their ability to discriminate the more subtle differences between genotypes and their interaction with environment, however, is less convincing (Lawn and Imrie, 1994). Some refinement of models to enable them to handle the details that determine genotype × environment interaction in a target region is thus necessary. Once this is done, application to breeding data to generate long-term "yield profiles" is likely to introduce some real advantages compared with conventional methodologies.

Several projects have illustrated how long-term crop yield "profiles" can be generated, and how data from such yield and other profiles can be used in decision making. Software packages that facilitate the generation and analysis of data in such a manner have also been made available in recent years (e.g., the DSSAT 3 package—Tsuji et al., 1995; APSIM—McCown et al., 1996). Results of the application of such a package to a rice-wheat cropping system were described by Timsina, Singh, and Singh (1997). Results of the application of the approach to wheat in Ontario, Canada, were reported by Hunt (unpublished data), who presented "performance profiles" over twenty years for winter wheat supplied with different amounts of N fertilizer at a site in Ontario, Canada (see Figure 10.4). The results indicate that decisions based on one three-year span of experimentation would be quite wrong for application to a different three-year period and emphasize either that experimentation for decision making should be continued for much longer than three years or that efforts to apply simulation models to extrapolate experimental data to a wider array of years should be enhanced.

For model use to generate long-term profiles of performance, values for the various genotype-specific coefficients used in the particular model chosen would be necessary. An essential precursor to model application in this context, therefore, is the development of software that makes use of one or more crop models to summarize trial data in terms of underlying genotypic aspects. This application reflects the fact that a simulation model is, in essence, a mechanism for combining genotypic characteristics with environmental parameters to produce estimates of crop performance, as expressed in the classic equation:

$$G \times E = P$$

where G, E, and P represent genotypic, environmental, and performance characteristics, respectively.

When "fed" with information on crop performance, such a model should be useable to produce estimates of genotypic characteristics.

FIGURE 10.4. Simulated Yields versus Year for Ruby Wheat, Elora Research Station, Ontario, Canada

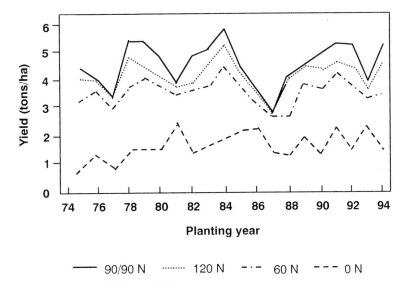

Planting year

——— 90/90 N 		········· 120 N 		– ·– 60 N 		– – – 0 N

Source: Hunt (unpublished data).

For example, a maize model could be used to deduce juvenile phase duration and photoperiod sensitivities from field data dealing with tasseling or silking dates, thus serving to resolve some of the genotype × environment interactions into discrete factors. Gencalc (Hunt et al., 1993) is an example of an initial attempt to develop such software. Further development of specific analytic software of this type is needed, however, before models are likely to be widely applied as analytic tools.

For analysis of data in this manner, customary trial or nursery data (e.g., heading date, grain yield) would have to be supplemented with a few additional plant measurements, and the environmental and soil data necessary to run the model would have to be collected. Crop data required, as a minimum, are listed in Table 10.4. Given such data, appropriate calibration software would run a model with initial guesses for various cultivar, and perhaps also soil, coefficients and then follow an iterative procedure until a set of coefficients that produces the best correspondence between model output and actual data over all the fully characterized trial or nursery sites is obtained. These cultivar coefficients would summarize the multilocation phenotypic data in terms of underlying genotype characteristics for the various cultivars in question.

CONCLUDING COMMENTS

It thus seems that crop simulation models could be used to increase the efficiency of crop improvement programs by facilitating the evaluation of yield potential, determination of yield gaps, or an analysis of performance data and an extrapolation of available data to many years. Use in these contexts, however, assumes confidence in the ability of the model to account for the various interactions in a realistic manner. Determining whether this confidence is justified will require much evaluation of model performance in comparison with data from focused field experiments. It is likely, however, that this will not be achieved until further detail has been included in the

TABLE 10.4. Crop Data Required for Computation of Genotype-Specific Characteristics of Maize

Emergence date
Date of silking
Date of maturity (black layer)
Date of onset of disease or insect attack
Number of leaves produced on the main axis
Grain yield (dry-weight basis)
Kernel dry weight
Harvest index
Grains per ear
Ear number per plant
Grain N percent
Damage from disease or insects

current array of crop models. Loomis (1993) has argued that models capable of simulating processes at a level more closely aligned to gene action are required. Others (Shorter, Lawn, and Hammer, 1991; Hammer et al., 1997) have argued that simpler crop physiological frameworks, which are more readily understood by crop physiologists and breeders, can be used and enhanced through the addition of more detail as this is shown to be necessary. This approach seems to be the most logical: one that, in essence, follows the scientific method. In following it, however, targeted experimentation that involves both breeders, those who are most acutely aware of genetic variation, and physiologists will be essential, together with careful model evaluation and improvement by crop physiologists and breeders, not just by modelers. Further, model structures must be such that additional detail can be added easily, without undermining the integrity of the model code and making the model virtually indecipherable to all but the originator. This has not always been the case in the past, and attention to structural aspects will, in many cases, be essential prior to the addition of more detail. Efforts along these lines are under way (Reynolds and Acock, 1997); they will need to be strengthened before widespread model use by crop physiologists and breeders is likely.

Once models of adequate power are available, and once these are coupled with appropriate support software, it is likely that they will be integrated with statistical procedures as the accepted mode of analysis of data in crop improvement. For the moment, use of models to provide "check" yield in breeding trials would seem to be the application most likely to be of value. Naturally, such application requires environmental documentation at trial sites. This is something that is receiving attention currently to permit better interpretation of data even when not using a model.

Beyond questions related to the integration of modeling with conventional breeding are many questions relating to the potential for integration with developments in genetic engineering. The potential offered by new molecular technologies, which are now being made use of in crop improvement programs, has, to date, revolved around single gene traits associated with pest and herbicide resistance or product quality. As technologies to deal with more complex traits are developed, the need to evaluate component processes in terms of their value in production environments will become greater. Modeling, by being able to supply comparative information on the value of subprocesses when these are operating in a whole crop system, will thus likely assume greater importance in crop improvement than it has today. All involved must therefore press forward in a concerted manner, so as to be in a position to meet these new demands when they are finally directed at us all.

REFERENCES

Aggarwal, P.K., Kropff, M.J., Teng, P.S., and Khush, G.S. (1997). The challenge of integrating systems approach in plant breeding: Opportunities, accomplishments and limitations. In Kropff, M.J., Teng, P.S., Aggarwal, P.K., Bouma, J., Bouman, B.A.M., Jones, J.W., and van Laar, H.H. (Eds.), *Applications of systems approaches at the field level,* Volume 2, (pp. 1-24). Dordrecht, Netherlands: Kluwer Academic Publishers.

Austin, R.B. (1975). The contribution of physiology to variety improvement. *Plant Breeding Institute, Annual Report 1975:* 58-63.

Boote, K.J., Jones, J.W., Hoogenboom, G., Wilkerson, G.G., and Jagtap, S.S. (1989). PNUTGRO V 1.02, *Peanut crop growth simulation model. User's guide. Florida Agricultural Experimental Station Journal* no. 8420. Gainesville, FL: University of Florida.

Boote, K.J., Jones, J.W., and Pickering, N.B. (1996). Potential uses and limitations of crop models. *Agronomy Journal* 88: 704-716.

Boote, K.J., Jones, J.W., and Singh, P. (1991). Modeling growth and yield of groundnut: State of the art. Proceedings of the International Workshop, *Groundnut: A global perspective* (pp. 331-343). Patancheru, AP, India: ICRISAT.

Carberry, P.S., Muchow, R.C., and McCown, R.L. (1993). A simulation model of kenaf for assisting fibre industry planning in northern Australia. 4. Analysis of climatic risk. *Australian Journal of Agricultural Research* 44: 713-730.

Cooper, M. and Hammer, G.L. (1997). *Plant adaptation and crop improvement.* Wallingford, UK: CAB International.

Crosbie, T.M. and Mock, J.J. (1981). Changes in physiological traits associated with grain yield improvement in three maize breeding programs. *Crop Science* 21: 255-259.

Dingkuhn, M., Jones, M.P., Fofana, B., and Sow, A. (1997). New high-yielding, weed competitive rice plant types drawing from *O. sativa* and *O. glaberrima* gene pools. In Kropff, M.J., Teng, P.S., Aggarwal, P.K., Bouma, J., Bouman, B.A.M., Jones, J.W., and van Laar H.H. (Eds.), *Applications of systems approaches at the field level,* Volume 2 (pp. 37-52). Dordrecht, Netherlands: Kluwer Academic Publishers.

Dingkuhn, M., Penning de Vries, F.W.T., DeDatta, S.K., and van Laar, H.H. (1991). New plant type concepts for direct seeded flooded rice. *Proceedings international rice research conference.* Seoul 1990. Los Baños, Philippines: International Rice Research Institute.

Donald, C.M. (1962). In search of yield. *Journal of the Australian Institute of Agricultural Science* 28: 171-178.

Donald, C.M. (1968a). The breeding of crop ideotypes. *Euphytica* 17: 385-403.

Donald, C.M. (1968b). The design of a wheat ideotype. In Finlay, K.W. and Shepherd, K.W. (Eds.), *Proceedings of the third international wheat genetic symposium* (pp. 377-387). Canberra, Australia: Australian Academy of Science.

Donald, C.M. (1979). A barley breeding programme based on an ideotype. *Journal of Agricultural Science, Cambridge* 93: 261-268.

Evans, L.T. (1993). *Crop evolution, adaptation and yield.* Cambridge: Cambridge University Press.

Field, T.R.O. and Hunt, L.A. (1974). The use of simulation techniques in the analysis of seasonal changes in the productivity of alfalfa (*Medicago sativa* L.) stands. In *Proceedings of the XII international grassland congress* (pp. 108-120). Moscow, USSR.

Frankel, H. (1947). The theory of plant breeding for yield. *Heredity* 1: 109-120.

Hammer, G.L., Butler, D.G., Muchow, R.C., and Meinke, H. (1997). Integrating physiological understanding and plant breeding via crop modelling and optimization. In Cooper, M. and Hammer, G.L. (Eds.), *Plant adaptation and crop improvement* (pp. 419-441). Wallingford, UK: CAB International.

Hunt, L.A. (1993). Designing improved plant types: A breeder's viewpoint. In Penning de Vries, F., Teng, P., and Metsellar, K. (Eds.), *Systems approaches for agricultural development* (pp. 3-18). Dordrecht, Netherlands: Kluwer Academic Publishers.

Hunt, L.A., Pararajasingham, S., Jones, J.W., Hoogenboom, G., Imamura, D.T., and Ogoshi, R.M. (1993). GENCALC: Software to facilitate the use of crop models for analyzing field experiments. *Agronomy Journal* 85: 1090-1094.

ICARDA (International Center for Agricultural Research in the Dry Areas) (1986). *Annual report 1985.* Aleppo, Syria: ICARDA, 378 pp.

Jensen, N.F. (1975). Breeding strategies for winter wheat improvement. In *Proceedings of the second international winter wheat conference* (pp. 31-45). Agricultural Institute, Zagreb, Yugoslavia.

Jones, J.W., Boote, K.J., Hoogenboom, G., Jagtap, S.S., and Wilkerson, G.G. (1989). SOYGRO V5.42, soybean crop growth simulation model. User's guide. *Florida Agricultural Experimental Station Journal* no. 8304. Gainesville, FL: University of Florida.

Jones, J.W. and Zur, B. (1982). Simulation of possible adaptive mechanisms in crop subjected to water stress. *Irrigation Science* 5: 251-264.

Jordan, W.R., Dugas, W.A., and Shouse, P.J. (1983). Strategies for crop improvement for drought-prone regions. *Agricultural Water Management* 7: 281-299.

Keating, B.A., Godwin, D.C., and Watiki, J.M. (1991). Optimising nitrogen inputs in response to climatic risk. In Muchow, R.C. and Bellamy, J.A. (Eds.), *Climatic* risk in crop production: Models and management for the semiarid tropics and subtropics* (pp. 329-358). Wallingford, UK: CAB International.

Lawn, R.J. and Imrie, B.C. (1991). Crop improvement for tropical and subtropical Australia: Designing plants for difficult climates. *Field Crops Research* 26: 113-139.

Lawn, R.J. and Imrie, B.C. (1994). Exploiting physiology in crop improvement. *Plant Physiology Abstracts* 20: 467-476.

Lindquist, J.L. and Kropff, M.J. (1997). Improving rice tolerance to barnyardgrass through early crop vigour: Simulations with INTERCOM. In Kropff, M.J., Teng, P.S., Aggarwal, P.K., Bouma, J., Bouman, B.A.M., Jones, J.W., and van Laar, H.H. (Eds.), *Applications of systems approaches at the field level,* Volume 2 (pp. 53-62). Dordrecht, Netherlands: Kluwer Academic Publishers.

Loomis, R.S. (1993). Optimization theory and crop improvement. In Buxton, D.R., Shibles, R., Forsberg, R.A., Blad, B.L., Asay, K.H., Paulsen, G.M., and Wilson, R.F. (Eds.), *International crop science,* Volume I (pp. 583-588). Madison, WI: Crop Science Society of America.

Marshall, D.R. (1991). Alternative approaches and perspectives in breeding for higher yields. *Field Crops Research* 26: 171-190.

McCown, R.L., Hammer, G.L., Hargreaves, J.N.G., Holzworth, D.P., and Fridman, D.M. (1996). APSIM: A novel software system for model development, model testing and simulation in agricultural system research. *Agricultural System* 50: 255-271.

Muchow, R.C. and Bellamy, J.A. (1991). *Climatic risk in crop production: Models and management for the semiarid tropics and subtropics.* Wallingford, UK: CAB International.

Muchow, R.C., Cooper, M., and Hammer, G.L. (1996). Characterizing environmental challenges using models. In Cooper, M. and Hammer, G.L. (Eds.), *Plant adaptation and crop improvement* (pp. 349-364). Wallingford, UK: CAB International.

Muchow, R.C., Hammer, G.L., and Carberry, P.S. (1991). Optimising crop and cultivar selection in response to climatic risk. In Muchow, R.C. and Bellamy, J.A. (Eds.), *Climatic risk in crop production: Models and management for the semiarid tropics and subtropics* (pp. 235-262). Wallingford, UK: CAB International.

Muchow, R.C. and Kropff, M.J. (1997). Assessing the potential yield of tropical crops: Role of field experimentation and simulation. In Kropff, M.J., Teng, P.S., Aggarwal, P.K., Bouma, J., Bouman, B.A.M., Jones, J.W., and van Laar, H.H. (Eds.), *Applications of systems approaches at the field level,* Volume 2 (pp. 101-112). Dordrecht, Netherlands: Kluwer Academic Publishers.

O'Toole, J.C. and Jones, C.A. (1987). Crop modeling: Applications in directing and optimizing rainfed rice research. In *Workshop on impact of weather parameters on the growth and yield of rice* (pp. 255-269). April 7-10, 1986. Manila, Philippines: International Rice Research Institute.

O'Toole, J.C. and Stockle, C.O. (1991). Conceptual and simulation modeling in plant breeding. In *Proceedings of the improvement and management of winter cereals under temperature, drought and salinity stresses symposium* (pp. 205-225). October 26-29, 1987. Cordoba and Madrid, Spain: INIA.

Penning de Vries, F.W.T. (1991). Development and use of crop modelling in rice research: Searching for higher yields. In *Proceedings of the symposium on rice research—New frontiers.* Hyderabad, India.

Reynolds, J.F. and Acock, B. (1997). Modularity and genericness in plant and ecosystem models. *Ecological Modelling* 94: 7-16.

Rickman, R.W., Waldman, S.E., and Klepper, B. (1996). MODWHT3: A development-driven wheat growth simulation. *Agronomy Journal* 88: 176-184.

Saulescu, N.N. and Kronstad, W.E. (1995). Growth simulation outputs for detection of differential cultivar response to environmental factors. *Crop Science* 35: 773-778.

Sharpley, A.N. and Williams, J.R. (1990). EPIC—Erosion/productivity impact calculator: 1. Model documentation. U.S. Department of Agriculture technical bulletin no. 1768, 235 pp.

Shorter, R., Lawn, R.J., and Hammer, G.L. (1991). Improving genotypic adaptation in crops—A role for breeders, physiologists and modellers. *Experimental Agriculture* 27: 155-175.

Sinclair, T.R. and Seligman, N.G. (1996). Crop modeling from infancy to maturity. *Agronomy Journal* 88: 698-704.

Singh, P., Boote, K.J., Rao, A.Y., Iruthayaraj, M.R., Sheikh, A.M., Hundal, S.S., Narang, R.S., and Singh, P. (1994). Evaluation of the groundnut model PNUT-GRO for crop response to water availability, sowing dates, and seasons. *Field Crops Research* 39: 147-162.

Singh, U. and Thornton, P.K. (1992). Using crop models for sustainability and environmental quality assessment. *Outlook on Agriculture* 21: 209-218.

Stockle, C.O. (1985). Simulation of the effect of water and nitrogen stress on growth and yield of spring wheat. PhD thesis. Washington State University, Pullman, WA.

ten Berge, H.F.M., Aggarwal, P.K., and Kropff, M.J. (1997). Applications of rice modelling. *Field Crops Research,* Special issue 51, 172 pp.

Timsina, J., Singh, U., and Singh, Y. (1997). Addressing sustainability of rice-wheat systems: Analysis of long-term experimentation and simulation. In Kropff, M.J., Teng, P.S., Aggarwal, P.K., Bouma, J., Bouman, B.A.M., Jones, J.W., and van Laar, H.H. (Eds.), *Applications of systems approaches at the field level,* Volume 2 (pp. 383-397). Dordrecht, Netherlands: Kluwer Academic Publishers.

Tsuji, G.Y., Jones, J.W., Hoogenboom, G., Hunt, L.A., and Thornton, P.K. (1995). Introduction. In Tusji, G.Y., Uehara, G., and Balas, S. (Eds.), *DSSAT Version 3,* Volume 1 (pp. 3-11). Honolulu, Hawaii: University of Hawaii.

Wagner-Riddle, C., Gillespie, T.J., Hunt, L.A., and Swanton, C.J. (1997). Modeling the effect of a rye cover crop and subsequent soybean yield. *Agronomy Journal* 89: 208-218.

Warren-Wilson, J. (1972). Control of crop processes. In Rees, A.R., Cockshull, K.E., Haul, D.W., and Hurd, R.G. (Eds.), *Crop processes in controlled environments* (pp. 7-30). London: Academic Press.

Willman, M.R., Below, F.E., Lambert, R.J., Howey, A.E., and Mies, D.W. (1987). Plant traits related to productivity of maize. I. Genetic variability, environmental variation, and correlation with grain yield and stalk lodging. *Crop Science* 27: 1116-1121.

Yan, W. and Hunt, L.A. (1998). Genotype by environment interaction and crop yield. In Janick, J. (Ed.), *Plant breeding reviews,* Volume 16 (pp. 135-178). New York: John Wiley and Sons, Inc.

Chapter 11

Recent Research
on Maize Grain Yield in Argentina

Gustavo A. Maddonni
Jorgelina Cárcova
María E. Otegui
Gustavo A. Slafer

INTRODUCTION

Maize grain yield is closely associated with final kernel number, which is related to plant growth rate at a critical period around tasseling/silking. As indicated by Andrade, Cirilo, and Echarte (see Chapter 5) and Westgate (see Chapter 7), much of the variation in kernel number could be explained through carbon assimilation and plant growth rate during this period, even for most (i.e., low to moderate) nutritional and water stress conditions met in field environments under which maize is produced in Argentina.

When water and nutrients are unlimited, differences in kernel number are related to (1) the amount of solar radiation intercepted during the critical period, which determines plant growth rate, and (2) genotypic differences in potential kernel number (Kiniry and Otegui, see Chapter 3). Under these conditions, limited opportunities can be expected for increasing kernel number through increased plant growth rate (Muchow, see Chapter 4). Attention should be given to temperature effects on crop cycle duration (Tollenaar and Wu, see Chapter 2; Muchow, see Chapter 4) and flowering dynamics (Edmeades, Bänziger, and Ribaut, see Chapter 6). Temperature may modify crop capacity to capture resources, while genotypic differences in flowering dynamics may affect pollination rate within the ear.

When water and nutrients are limiting, strategies for maximizing crop production can be addressed by adopting agronomic practices aimed at avoiding (e.g., sowing date, fertilization) or partially alleviating (e.g., plant population) the stress effects, together with the use of hybrids of different rates of development and with traits that promote stress tolerance (Edmeades, Bänziger, and Ribaut, see Chapter 6).

In this chapter, we summarize new research being conducted in Argentina on maize grain yield determination that may be valuable for maintaining or increasing current genotypic gains in maize yield. Information was extracted from the poster session of the workshop and organized in three sections: (1) breeding effects on maize kernel number determination, (2) pollination synchrony and maize kernel number, and (3) water deficit.

BREEDING EFFECTS ON MAIZE KERNEL NUMBER DETERMINATION

Echarte and colleagues (1998) analyzed the response of kernel number per plant (KNP) to plant growth rate (PGR) during the critical period for a set of seven hybrids released between 1965 and 1995 in Argentina. Changes in PGR were achieved through varying plant population between 5 and 18.5 plants per square meter (plants/m^2). The relationship between KNP and PGR was always described by a double inverse function, each part of the type:

$$KNP = a + b/PGR \qquad (11.1)$$

with the second curve representing the contribution of the subapical ear to final kernel number (see Figure 5.4, in Chapter 5). Only one curve appears when a nonprolific hybrid is analyzed (see Figure 11.1a), whereas two curves appear when the hybrid × environment condition allows a second ear to grow and set kernels.

Modern hybrids exhibited a greater KNP than the old ones, mainly due to genetic improvement in potential kernel number per uppermost ear (parameter a in equation 11.1; see Figure 11.1b). With decreasing PGR, modern hybrids were also able to set more kernels than the old ones, and the estimate of the initial slope of the relationship between KNP and PGR (parameter *b* in equation 11.1) was larger in modern than in old cultivars as well (see Figure 11.1c). All hybrids showed positive abscissa intercepts, which are indicating PGR thresholds for kernel set, in agreement with Tollenaar, Dwyer, and Stewart (1992). Hybrids did not show any trend to reduce the minimum PGR to set kernels along the three decades analyzed (see Figure 11.1d), nor were clear trends detected on prolificacy (Echarte et al., 1998). Most of the differences in KNP produced by hybrid breeding in Argentina appeared, then, to be the consequence of associated changes in maximum KNP and the ability to set more kernels at very low PGR (see Figure 11.1e), with negligible differences in the threshold PGR for setting kernels (see Figure 11.1d).

The increased efficiency to set kernels of the new materials has also been related to an enhanced anthesis-silking synchrony and a more synchronous pollination within the ear (Luque et al., 1998). Luque and colleages (1998) analyzed the effects of breeding on floral synchrony of three maize hybrids released in Argentina in 1965, 1985, and 1993. In their study, plant population was varied between 3 and 18 plants/m^2, and the number of plants shedding pollen and showing newly emerged silks was monitored every two to three days (i.e., the number of days with anthers shedding pollen and ears exerting new silks was recorded for each plant). Their data revealed that breeding has shortened the average anthesis-silking interval (ASI) of commercial hybrids, as has also been the case in a maize population breeding program for droughted conditions (Edmeades, Bänziger, and Ribaut, see Chapter 6). Averaged across plant populations, the most modern hybrid had a protandry (days between 50 percent of the plants shedding pollen and 50 percent of the plants with visible silks) of 1.4 days and the oldest, 3.6 days (see Figure 11.2). This change in protandry was not related to a reduction in time to flowering, which was shorter for the old hybrid than for the new one (Luque et al., 1998). The most remarkable difference, however, was not in their averaged ASIs, but in the response of this trait to plant population (see Figure 11.2). No difference was apparent at relatively low populations, but as competition increased, the newest hybrid exhibited a strong stability in ASI, allowing silking to occur slightly after pollen shedding, while the oldest hybrid increased its ASI steadily with plant population increase (see Figure 11.2). Thus,

FIGURE 11.1. Schematic Response of Kernel Number per Apical Ear to Plant Growth Rate Using an Inverse Function

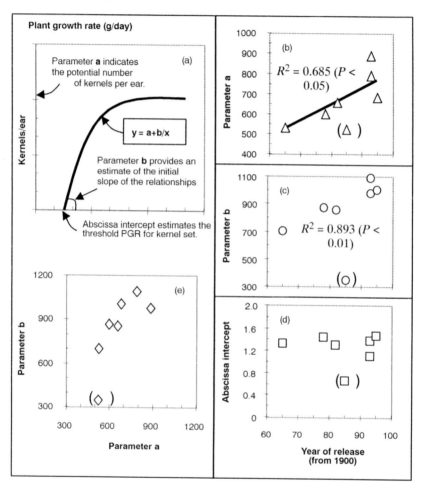

Source: Calculated from Echarte et al. (1998).

Note: (a) Response of the parameters of the inverse function (figures b and c) and the calculated abscissa intercept (d) to the year of release, and relationship between both parameters of the inverse function (e). Data in parentheses correspond to *prolific* hybrid DK 4F37.

it is likely that the positive trend of KNP with year of release observed by Echarte and colleagues (1998; see Figure 11.1), which was at least partly due to higher kernel set at very low PGRs (i.e., very high plant populations), had been associated with the ability of modern hybrids to maintain a certain degree of synchrony at very low PGRs (see Figure 11.2).

POLLINATION SYNCHRONY AND MAIZE KERNEL NUMBER

To further explore the direct impact a more synchronous pollination may have on final kernel number, Cárcova and colleagues (1998) performed two pollination treatments (natural pollination and hand pollination) on both the uppermost and subapical ears of two

FIGURE 11.2. Relationship Between the Anthesis-Silking Interval and Plant Density for an Old (Triangle) and a New Hybrid (Circle), Released in Argentina in 1965 and 1985, Respectively

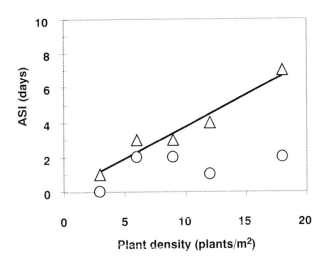

Source: Raw data taken or calculated from Luque et al. (1998).

Note: Line fitted by regression for the old hybrid ($R^2 = 0.95$).

hybrids grown at two plant populations. Hybrids differed in prolificacy type (i.e., ears per plant) and potential kernel number (i.e., spikelets per ear). Ears (apical and subapical) of hand-pollinated plants remained covered up to silking, plus six days, when all exposed silks were pollinated simultaneously, while controls were left uncovered and pollinated naturally.

Synchronous hand pollination increased kernel set with respect to natural pollination in all treatments. For both hybrids, differences in kernel number between pollination treatments were related to

- more kernels in the apical ear at high plant population (9 plants/m^2) and
- more kernels from subapical ears at the low plant population (3 plants/m^2).

No subapical grain-bearing ear was registered at the high plant population for any hybrid and pollination treatment, but hand pollination increased prolificacy of the nonprolific hybrid at the low plant population.

Data obtained in this research agreed with results on grain yield per plant (Sarquís, Gonzalez, and Dunlap, 1998) and kernel set per ear (see Figure 7.2, Chapter 7) reported in studies on pollination control and may partially explain the positive effects of a reduced ASI on final kernel number and grain yield (discussed in the previous section and in Chapters 6 and 7). Cárcova and co-authors (1998) concluded that synchronous pollination improved final kernel number and partially offset the negative effect of the silking interval between ears on kernel set in the subapical ear. They suggested that for environments in which grain yield can vary markedly depending on rainfall around silking, and low plant populations are used to partially compensate this constraint, selection should focus on hybrids capable of setting kernels in the subapical ear. This trait is strongly dependent on the silking interval between ears but can also be improved via artificially managed synchronous pollination (e.g., male-sterile hybrid combined with a pollinator hybrid).

WATER DEFICIT

Information on drought tolerance of corn hybrids can be inferred from the stability index (Eberhart and Russell, 1966), which has been currently expressed as the slope of the linear relationship between grain yield and the environment index (i.e., estimated as grain yield averaged across all hybrids in a given environment; Finlay and Wilkinson, 1963). Few studies, however, attempted to establish the relationship between the stability index (*b*, stability parameter) and those processes or plant characteristics (e.g., osmotic adjustment, rooting capacity, flowering dynamics) with possible effects for improving grain production in plant breeding programs (Edmeades, Bänziger, and Ribaut, see Chapter 6).

Osmotic Adjustment

Lemcoff, Chimanti, and Davezac (1998) demonstrated the existence of genetic variation in osmotic adjustment for nine Argentine hybrids. Differences in this trait were positively related to those in stability index (see Figure 11.3), which ranged between 0.89 and 1.25. These authors suggested that selection for osmotic adjustment can be performed in early vegetative stages, producing, as a result, concomitant differences in osmotic adjustment at silking.

Rooting Pattern and Canopy Conductance

Recent studies in the Rolling Pampas (32° to 35°S, 58° to 62°W), which is the main maize growing region in Argentina, have established that long-term conventional cropping (e.g., based on the use of the moldboard plow) caused physical and chemical degradation of soil properties (Maddonni, Urricariet, et al., 1999), resulting in reduced water availability for maize crops. Nevertheless, permanent adoption of modern hybrids, which outyield the old ones even under drought conditions, has partially hidden increased soil degradation, resulting in a steady growth of maize yield during recent decades (Slafer and Otegui, see Chapter 1), despite the negative effects of intensive land use.

Cárcova and Maddonni (1998) analyzed canopy conductance and the rooting pattern (i.e., root length density) as possible plant traits

FIGURE 11.3. Stability Parameter (*b*) and Relative Water Content (RWC$_{-1.64}$) of the Uppermost Expanded Lamina at -1.64 MPa of Osmotic Potential

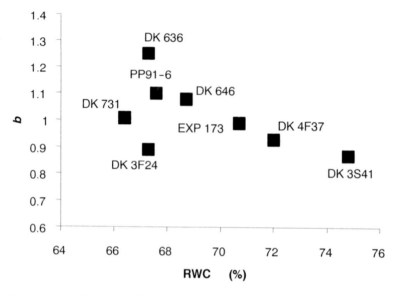

Source: Calculated from Tables 1 and 2 in Lemcoff, Chimenti, and Davezac (1998).

Note: Data are accompanied by hybrid numbers. Water withheld during a thirty-day period, starting at forty-three days after emergence.

conferring differential water extraction capacity to three maize hybrids released in 1975 (DK 4F31), 1987 (DK 4F37), and 1993 (DK 752). Only a small difference between hybrids emerged in crop evapotranspiration (ET) at silking in one of the three environments analyzed. In this case, the hybrid DK 4F37 showed the highest water extraction capacity, which was related to its high canopy conductance value (see Table 11.1). These characteristics were in agreement with the high osmotic adjustment described for this hybrid (Lemcoff et al., 1998). Hybrids differed in the pattern of root distribution, with the most common pattern corresponding to the modern hybrid DK 752 (see Figure 11.4). However, its more uniform rooting pattern did not represent an advantage in water extraction (see Table 11.1).

TABLE 11.1. Crop Evapotranspiration (ET) and Canopy Conductance (ET/IPAR) Around Silking of Three Hybrids (DK 4F31, DK 4F37, and DK 752) Grown in Soils with Short- and Long-Term Cropping in Two Sites (1994-1995) of the Rolling Pampas (Farm 1 and Farm 3)

	1994-1995		1995-1996	
	ET mm/day	ET/IPAR mm $(MJ \cdot m^{-2})^{-1}$	ET mm/day	ET/IPAR mm $(MJ \cdot m^{-2})^{-1}$
Farm 1				
Short term	5.1	0.70	3.3	0.43
Long term	3.8	0.57	2.4	0.30
	(0.001)	(0.08)	(0.1)	ns
DK 4F31	4.5 b	0.64 ab	2.8	0.41 b
DK 4F37	4.7 a	0.71 a	2.9	0.45 a
DK 752	4.2 c	0.56 b	2.7	0.39 b
	(0.001)	(0.05)	ns	(0.05)
Farm 3				
Short term	5.5	0.72		
Long term	4.5	0.62		
	(0.05)	(0.10)		
DK 4F31	5.0	0.67		
DK 4F37	5.0	0.70		
DK 752	5.0	0.66		
	ns	ns		
ET_m	5.7		5.9	

Source: Cárcova and Maddonni (1998).

Note: Data from 1995-1996 are the average of three sites. Values in parentheses indicate the significance level of the corresponding ANOVA, while different letters indicate significant differences between treatment means evaluated within each ANOVA. ET_m represents potential evapotranspiration of the same period.

High water extraction around silking and canopy conductance, however, were not necessarily reflected in improved final kernel number. This result indicates that attributes which confer improved performance under low and moderate stress might be constitutive rather than stress related (Blum, 1997), and that selection for higher yield potential may result in the release of hybrids with improved performance under water stress (Slafer and Araus, 1998). In the study by Cárcova and Maddonni (1998), the most recently released hybrid, which tended to exhibit smaller values for water extraction

FIGURE 11.4. Root Length Density Between 0 and 80 cm Soil Depth of Three Hybrids (DK 4F31, DK 4F37, and DK 752)

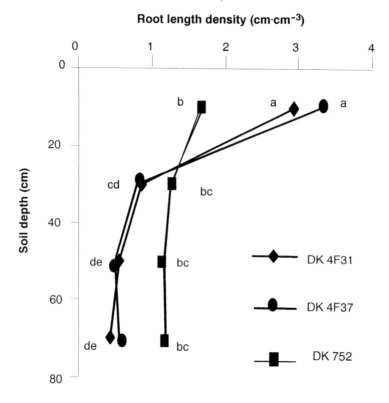

Source: Cárcova and Maddonni (1998).

Note: Different letters at each layer indicate significant differences ($P < 0.05$) between hybrids.

and canopy conductance, presented the highest kernel number per millimeter (mm) of ET_a, particularly for ET_a greater than 3 mm/day (see the next section).

Flowering Dynamics

Results observed by Cárcova and Maddonni (1998) could be partially explained by differences in silking rate among hybrids. Maddonni and Cárcova (1998) determined that the hybrid which exposed more silks synchronously for pollination (DK 752, see Table 11.2) had an increased kernel set per mm of evapotranspired water (see Figure 11.5). Hence, this hybrid exhibited the largest number of kernels per square meter and yield (data not shown) in most tested environments (one of which was stress free).

These results, obtained with a similar set of hybrids to those used by Echarte and colleagues (1998) and Luque and colleagues (1998), reinforced the importance of flowering dynamics and pollination synchrony with regard to kernel set and grain yield (see Chapters 6 and 7) and expand the importance of this trait for improving yield under conditions of water stress.

TABLE 11.2. Tasseling and Silking (Apical Ear) Progress Rates (Relative Increment per 100 °Cdays) of Three Hybrids Grown in a Control Experiment and in Soils with Short- and Long-Term Cropping Histories During 1995-1996

| | Tasseling progress rate | | | | Silking progress rate | | | |
| | | Cropping history | | | | Cropping history | | |
Hybrid	Control	Short	Long	Mean	Control	Short	Long	Mean
DK 4F31	0.93	0.16 b	0.39 a	0.27 b	0.69	0.54 b	0.33 a	0.43 ab
DK 4F37	0.76	0.62 a	0.48 a	0.55 a	0.89	0.51 b	0.14 a	0.33 b
DK 752	1.55	0.57 a	0.62 a	0.60 a	1.02	0.88 a	0.22 a	0.55 a
Mean	1.08	0.45	0.50		0.87	0.64 A	0.23 B	

Source: Maddonni and Cárcova (1998) and Maddonni, Cárcova, et al. (1999).

Notes: Different lowercase letters within a column indicate significant ($P < 0.05$) differences between hybrids. Different capital letters within a row indicate significant ($P < 0.05$) differences between cropping histories.

FIGURE 11.5. Kernel Number and Crop Evapotranspiration (ET) Around Silking

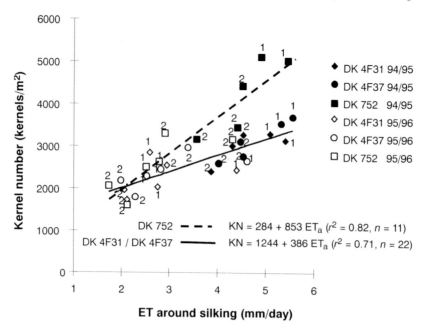

Source: Cárcova and Maddonni (1998).

Note: 1 and 2 stand for extreme cases of soil degradation: 1: short cropping history; 2: long cropping history.

CONCLUSIONS

From recent research conducted in Argentina, the most relevant ideas to be considered in maize breeding are summarized in the following points:

- Genetic improvement of potential kernel number per ear together with an increased capacity to set kernels at low PGR have promoted increased final KNP of modern hybrids with respect to old maize hybrids. On the other hand, modern hybrids did not show any trend to reduce the minimum PGR to set kernels.

- Increased efficiency (i.e., kernel set per unit PGR) of new materials to set kernels has been related to an enhanced anthesis-silking synchrony and a more synchronous pollination within the ear. The most remarkable difference between modern and old hybrids is not related to their averaged ASIs, but to the repsonse of this trait to plant population increase. As competition was intensified, new hybrids exhibited a stronger stability in their ASI values than old hybrids. This trait might be responsible for the enhanced kernel set of new hybrids at low PGRs. This hypothesis is supported by data obtained in experiments where pollination was controlled.
- The importance of flowering dynamics on kernel set was reinforced by data obtained under water stress conditions. In these environments, traits such as the rooting pattern and canopy conductance, which exhibited genotypic variability, were not reflected in final kernel number improvement. Apparently, attributes that confer improved performance under moderate stress might be constitutive rather than stress related.

REFERENCES

Blum, A. (1997). Constitutive traits affecting plant performance under stress. In Edmeades, G.O., Bänziger, M., Mickelson, H.R., and Peña-Valdivia, C.B. (Eds.), *Developing drought- and low-N-tolerant maize* (pp. 136-141). Mexico DF, Mexico: CIMMYT.

Cárcova, J., Borrás, L., Uribelarrea, M., and Otegui, M.E. (1998). Synchronous pollination improves kernel set in maize. In Otegui, M.E. and Slafer, G.A. (Eds.), *International workshop on physiological bases for maize improvement* (pp. 110-111). Buenos Aires, Argentina: Impresos Agronomía.

Cárcova, J. and Maddonni, G.A. (1998). Performance of maize hybrids grown in soils with contrasting agricultural history. II. Kernel set. In Otegui, M.E. and Slafer, G.A. (Eds.), *International workshop on physiological bases for maize improvement* (pp. 116-117). Buenos Aires, Argentina: Impresos Agronomía.

Eberhart, S.A. and Russell, W.A. (1966). Stability parameters in comparing varieties. *Crop Science* 6: 36-40.

Echarte, L., Vega, C., Andrade, F.H., and Uhart, S.A. (1998). Kernel number determination in Argentine maize hybrids released during the last three decades. In Otegui, M.E. and Slafer, G.A. (Eds.), *International workshop on physiological bases for maize improvement* (pp. 102-103). Buenos Aires, Argentina: Impresos Agronomía.

Finlay, K.W. and Wilkinson, G.N. (1963). The analysis of adaptation in a plant breeding program. *Australian Journal of Agricultural Research* 14: 742-754.

Lemcoff, J.H., Chimenti, C.A., and Davezac, T.A.E. (1998). Osmotic adjustment in maize (*Zea mays* L.): Changes with ontogeny and its relationship with phenotypic stability. *Journal of Agronomy and Crop Science* 180: 241-247.

Luque, S.F., Cirilo, A.G., Otegui, M.E., and Andrade, F.H. (1998). Floral synchrony in maize: Changes introduced with genetic improvement in Argentina. In Otegui, M.E. and Slafer, G.A. (Eds.), *International workshop on physiological bases for maize improvement* (pp. 104-105). Buenos Aires, Argentina: Impresos Agronomía.

Maddonni, G.A. and Cárcova, J. (1998). Performance of maize hybrids grown in soils with contrasting agricultural history. I. Tasseling and silking progress. In Otegui, M.E. and Slafer, G.A. (Eds.), *International workshop on physiological bases for maize improvement* (pp. 114-115). Buenos Aires, Argentina: Impresos Agronomía.

Maddonni, G.A., Cárcova, J., Iglesias Pérez, M.E., and Ghersa, C.M. (1999). Maize flowering dynamic in soils with contrasting agricultural history. *Maydica* 44: 141-147.

Maddonni, G.A., Urricariet, S., Ghersa, C.M., and Lavado, R.S. (1999). Assessing soil quality in the Rolling Pampa using soil properties and maize characteristics. *Agronomy Journal* 91: 280-287.

Sarquís, J.I., Gonzalez, H., and Dunlap, J.R. (1998). Yield response of two cycles of selection from a semiprolific early maize (*Zea mays* L.) population to plant density, sucrose infusion, and pollination control. *Field Crops Research* 55: 109-116.

Slafer, G.A. and Araus, J.L. (1998). Improving wheat responses to abiotic stresses. In Slinkard, A.E. (Ed.), *Keynote addresses and oral presentations of the proceedings of the 9th international wheat genetics symposium* (pp. 201-213). University of Saskatchewan, Canada: University Extension Press.

Tollenaar, M., Dwyer, L.M., and Stewart, D.W. (1992). Ear and kernel formation in maize hybrids representing three decades of grain yield improvement in Ontario. *Crop Science* 32: 432-438.

Index

Page numbers followed by the letter "f" indicate figures; those followed by the letter "t" indicate tables.

Order Your Own Copy of
This Important Book for Your Personal Library!

PHYSIOLOGICAL BASES FOR MAIZE IMPROVEMENT

_____ in hardbound at $69.95 (ISBN: 1-56022-889-X)

COST OF BOOKS_____

OUTSIDE USA/CANADA/
MEXICO: ADD 20%_____

POSTAGE & HANDLING_____
(US: $4.00 for first book & $1.50
for each additional book
Outside US: $5.00 for first book
& $2.00 for each additional book)

SUBTOTAL_____

IN CANADA: ADD 7% GST_____

STATE TAX_____
(NY, OH & MN residents, please
add appropriate local sales tax)

FINAL TOTAL_____
(If paying in Canadian funds,
convert using the current
exchange rate. UNESCO
coupons welcome.)

☐ **BILL ME LATER:** ($5 service charge will be added)
(Bill-me option is good on US/Canada/Mexico orders only;
not good to jobbers, wholesalers, or subscription agencies.)

☐ Check here if billing address is different from
shipping address and attach purchase order and
billing address information.

Signature_____

☐ **PAYMENT ENCLOSED:** $_____

☐ **PLEASE CHARGE TO MY CREDIT CARD.**

☐ Visa   ☐ MasterCard   ☐ AmEx   ☐ Discover
☐ Diner's Club   ☐ Eurocard   ☐ JCB

Account # _____

Exp. Date _____

Signature _____

Prices in US dollars and subject to change without notice.

NAME _____
INSTITUTION _____
ADDRESS _____
CITY _____
STATE/ZIP _____
COUNTRY _____ COUNTY (NY residents only) _____
TEL _____ FAX _____
E-MAIL_____

May we use your e-mail address for confirmations and other types of information? ☐ Yes   ☐ No
We appreciate receiving your e-mail address and fax number. Haworth would like to e-mail or fax special
discount offers to you, as a preferred customer. **We will never share, rent, or exchange your e-mail
address or fax number.** We regard such actions as an invasion of your privacy.

Order From Your Local Bookstore or Directly From

The Haworth Press, Inc.

10 Alice Street, Binghamton, New York 13904-1580 • USA

TELEPHONE: 1-800-HAWORTH (1-800-429-6784) / Outside US/Canada: (607) 722-5857

FAX: 1-800-895-0582 / Outside US/Canada: (607) 772-6362

E-mail: getinfo@haworthpressinc.com

PLEASE PHOTOCOPY THIS FORM FOR YOUR PERSONAL USE.

www.HaworthPress.com

BOF00